NEW DEVELOPMENTS IN MEDICAL RESEARCH

AN INTRODUCTION TO ANTIBACTERIAL PROPERTIES

NEW DEVELOPMENTS IN MEDICAL RESEARCH

Additional books and e-books in this series can be found on Nova's website under the Series tab.

NEW DEVELOPMENTS IN MEDICAL RESEARCH

AN INTRODUCTION TO ANTIBACTERIAL PROPERTIES

NICHOLAS PAQUETTE
EDITOR

nova
Medicine & Health
New York

Library of Congress Cataloging-in-Publication Data

ISBN: 978-1-53618-305-4
Names: Paquette, Nicholas, editor.
Title: An introduction to antibacterial properties / Nicholas Paquette, editor.
Description: New York : Nova Science Publishers, [2020] | Series: New developments in medical research | Includes bibliographical references and index. |
Identifiers: LCCN 2020030191 (print) | LCCN 2020030192 (ebook) | ISBN 9781536183054 (paperback) | ISBN 9781536183054 (adobe pdf)
Subjects: LCSH: Drug resistance in microorganisms. | Antibiotics--Development.
Classification: LCC QR177 .A48 2020 (print) | LCC QR177 (ebook) | DDC 616.9/041--dc23
LC record available at https://lccn.loc.gov/2020030191
LC ebook record available at https://lccn.loc.gov/2020030192

Published by Nova Science Publishers, Inc. † New York

CONTENTS

PREFACE

An Introduction to Antibacterial Properties provides an overview of the current antibacterial techniques and main strategies being employed for controlling the colonisation of bacteria on surfaces, together with the design of antibacterial surfaces and their fabrication techniques.

The authors discuss the characteristic mechanisms of five main aspects of antibacterial agents and the corresponding representative type of antibacterial agents.

In addition, recent findings in (meth)acrylate hydrogels containing metal ions with antimicrobial properties are summarized.

Antibiotics, commonly referred to as "wonder drugs", have been used for various therapeutic purposes for decades. As such, this compilation addresses a concomitant problem: the emergence of drug resistance.

Antimicrobial adjuvants can aid in the prevention of antimicrobial resistance by suppressing the emergence of bacterial resistance and rejuvenating the antimicrobial activity of currently available commercial antibiotics cost-effectively.

Chapter 1 - An Introduction to Antibacterial Properties provides an overview of the current antibacterial techniques and main strategies being employed for controlling the colonisation of bacteria on surfaces, together with the design of antibacterial surfaces and their fabrication techniques.

The authors discuss the characteristic mechanisms of five main aspects of antibacterial agents and the corresponding representative type of antibacterial agents.

In addition, recent findings in (meth)acrylate hydrogels containing metal ions with antimicrobial properties are summarized.

Antibiotics, commonly referred to as "wonder drugs", have been used for various therapeutic purposes for decades. As such, this compilation addresses a concomitant problem: the emergence of drug resistance.

Antimicrobial adjuvants can aid in the prevention of antimicrobial resistance by suppressing the emergence of bacterial resistance and rejuvenating the antimicrobial activity of currently available commercial antibiotics cost-effectively.

Chapter 2 - Antibacterial agent is a natural or synthetic compound which inhibits the growth of bacteria in microorganisms. These agents determine their antibacterial range according to the type of targeted bacteria and are commonly divided into gram-positive bacteria and gram-negative bacteria. If agents act on both gram-positive and -negative bacteria, it can be classified as a broad-spectrum antibacterial agent. Moreover, if only a few of them work on it, they can be classified as a medium-spectrum antibacterial agent. In general, antibacterial agents are categorized based on structural features and mode of action. These include five main topics: inhibition of cell wall synthesis, disturbance of nucleic acid synthesis, protein synthesis, folic acid and cell membrane dysfunction by changing cell membrane permeability. This chapter focused on characteristic mechanism of five main categories and the corresponding representative type of antibacterial agents.

Chapter 3 - The rapid emergence of antibiotic-resistant pathogens is becoming an imminent global public health problem. The development of novel antimicrobial materials aiming to prevent or control infections caused by these pathogens is a very important issue. Polymeric hydrogels are versatile materials, which can be a great alternative to conventional treatments of infections. Because of its high hydrophilicity, unique three-dimensional network, fine biocompatibility, and cell adhesion, the hydrogels are suitable biomaterials for drug delivery in antimicrobial areas. The biocompatible nature of hydrogels makes them a convenient starting platform to develop selectively active antimicrobial materials. Hydrogels with antimicrobial properties can be designed by loading of known antimicrobial agents, or the material itself can be designed to possess inherent antimicrobial activity. The combination of polymeric hydrogels based on (meth)acrylate with metal ions (Ag, Cu, and Zn) is a simple and effective approach for obtaining multicomponent systems with diverse functionalities. Silver (Ag^{+}), copper (Cu^{2+}), and zinc (Zn^{2+}) ions have been loaded into hydrogels for antimicrobial applications. The incorporation of metal ions into hydrogels not only enhances the antimicrobial activity of hydrogels but also influences their swelling and release characteristics. Herein, the authors summarize recent findings in (meth)acrylate hydrogels containing metal ions with antimicrobial properties.

Chapter 4 - Antibiotics, addressed as the 'wonder drugs', are used for various therapeutic purposes for decades and a concomitant problem is the emergence of drug resistance, which is a challenging issue before the scientific community and the pharmaceutical industry.

Understanding the various aspects of the use of antibiotics, like targets of antibiotics, viz. cell wall synthesis, protein synthesis, nucleic acid synthesis and folate synthesis, the antibiotics used as inhibitors and their mechanisms of action, i.e., prevention of access to drug targets, changes in the structure and protection of antibiotic targets and the direct modification or inactivation of antibiotics, the emergence and causes of antibiotic resistance and strategies to overcome the antibiotic resistance are the focal points discussed in this chapter.

The antibiotic resistance crisis has been attributed to overuse and abuse of antibiotics, excessive prescribing, and slackness in new drug development by the pharmaceutical industry due to reduced economic incentives and daunting regulatory requirements. Designing novel analogs, drug combinations, innovative therapeutic approaches, and developing antibiotics with new mechanisms are considered to be the most effective strategies to combat the inevitable phenomena of drug resistance.

Chapter 5 - Antibacterial resistance is currently a global challenge since the number of resistant strains against multiple antibiotics continuously increasing, and there is an urgent need to develop novel strategies to overcome this problem. The outer protective membrane and vital overexpressed efflux pumps are some key factors responsible for intrinsic resistance in Gram-negative bacteria.

The antibiotic drug discovery process comprising chemical modification of existing antibiotic scaffolds has been proven very successful to expand potency, spectrum, and bypass resistance pathways. However, the emergence of drug-resistant microbial strain severely compromised the effectiveness of currently available commercial antibiotics. The antimicrobial resistance is the result of evolution and an unavoidable condition, which leads to the conclusion that developing an alternative perspective for treatment options is vital. Though, many strategies may be employed to minimize the impact and emergence of resistance, among them combination therapies are one of such effective strategies specially adjuvants that are chemically active moieties with no antibiotic action. Antimicrobial adjuvants acted as a blocker of antimicrobial resistance or booster of antimicrobial action. This approach not only suppresses the emergence of bacterial resistance but rejuvenate the antimicrobial activity of currently available commercial antibiotics cost-effectively.

In: An Introduction to Antibacterial Properties ISBN: 978-1-53618-305-4
Editor: Nicholas Paquette

Chapter 1

AN INTRODUCTION TO ANTIBACTERIAL PROPERTIES

Xiaochao Ji[1,*], Yue Kang[1], Tian Ma[1] and Helong Yu[2,†]
[1]Institute of Quartmaster Engineering and Technology,
Beijing, China
[2]National Key Laboratory for Remanifacturing,
Beijing, China

ABSTRACT

The emergence of some superbugs makes Hospital-Acquired Infections (HAIs) one of the chief causes of death all around the world. It is important to develop effective preventative technology to reduce bacterial adherence to the medical devices. A variety of approaches have been developed for the construction of biomaterials and functional surfaces with antibacterial properties in the past years. The type of antibacterial strategy is usually determined by the potentially adhering bacteria and bactericidal activity, which is affected by morphology, environment conditions, bacteria features, and antibacterial agents. The antibacterial

* Corresponding Author's Email: jixiaochao@gmail.com.
† Corresponding Author's Email: helong.yu@163.com.

mechanisms can be subjected to chemical or physical approaches. The complex interactions between the bacteria cell walls and the antibacterial agents have been studied by researchers to developed advanced antibacterial surfaces. This chapter will provide an overview of the current developed antibacterial techniques and main strategies being employed for controlling the colonisation of bacteria on surfaces, together with the design of antibacterial surfaces and their fabrication techniques.

Keywords: antibacterial surface, antiadhesive surface, nanostructure, bactericidal, coating

INTRODUCTION

Bacteria have existed on our planet hundred millions years, which can be found almost everywhere on Earth, and have developed excellent adaptive ability. Bacteria live in symbiotic and parasitic relationships with flora and fauna. The majority of the bacteria in body are harmless by the protective effects of immune system, while some bacteria are pathogenic and can cause serious diseases. Humans have fought against the detrimental effects caused by pathogenic bacteria throughout history (Lewis and Shan 2017). The bubonic plague known as 'Black Death' is famous for killing about 25 million people in the fourteenth century, which was caused from human beings infected by a bacterium called *Yersinia pestis.* According to the data from World Health Organization (WHO), tuberculosis caused by *Mycobacterium tuberculosis* alone infected 8.6 million people and leading to 1.3 million death in 2012. In addition, hospital acquired infections (HAIs) caused by bacteria such as methlicillin-resistant *Staphylococcus aureus* (MRSA) and *Pseudomonas aeruginosa* are ranked in the top 10 causes of death (Tripathy et al. 2017). Thus, bacteria infection is a global healthcare challenge and tremendous efforts have been made to solve the problem. The number of antibacterial related publications is summurized in Figure 1, and it can be seen that increasing attentions have been received in this field in order to develop advanced antibacterial techniques.

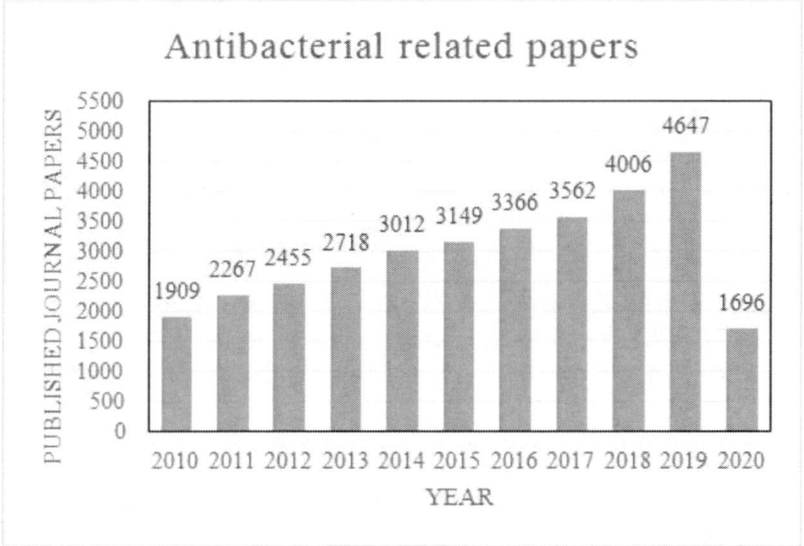

Figure 1. Trends of publications on antibacterial related papers (Data from Web of Science, 7/5/2020).

Bacterial adhesion on the surfaces of various exteriors like cell phones, food packages, surgery tools, air-conditioning system, etc. can cause serious problems in our society from the perspective of economical and health (Swartjes et al. 2015). Beside pathogenesis and diseases, adhesion of bacteria can cause other problems, such as spoilage, blockage, or reduction of energy transfer efficiency. For example, Food spoilages caused by bacteria are common in the food industry, and various strategies have been developed to extend the preserving time. Large amount of expenses was cost in cleaning bacteria attached to the ship hulls, which can reduce the fuel efficiency of ships. It is also reported that the bacteria can cause biocorrosion, and pits will form at the bacteria attached surfaces, accelerating the corrosion process. Thus, it is important to control bacterial adhesion. However, bacteria have high metabolic and physiological adaptability, which enables them to tune their genetic features to colonize in harsh environments.

Dense communities formed by attached bacteria are called biofilms, which is an effective way for bacteria to grow and proliferate. A biofilm can be formed by single specie or multiple species of bacteria. These aggregations are beneficial for bacteria survival because the community can enhance bacteria resistance to antibiotics and disinfectants, and enhance the regulation of gene expression. Besides, channel networks within the biofilms enable better diffusion of nutrients. The formation of biofilm begins with the colonization of a surface by bacteria. After adhesion, bacteria start to grow and divide, and then the density of the community increase. The extracellular polymeric substance (EPS) will be excreted by bacteria when a mature community is formed. Channels will be generated within the community for the intake of nutrients and the expulsion of wastes, whilst biofilms provide long periods for the bottom bacteria to develop antibiotic resistance to the specific antibiotic agents. To start a new circle of biofilm formation, some individual bacteria will be released from the biofilm and spread to colonize new surfaces. Biofilms have excellent mechanical and chemical resistance to a harsh environment, and once a biofilm formed, it is considerably difficult to tackle the bacteria colonies. Thus, it is essential to prevent the formation of biofilms by effectively killing the bacteria once they contact the surface (Lin et al. 2017).

Intensive efforts have been made to prevent the contamination of bacteria throughout our history. Inorganic antibacterial agents, such as silver, copper, zinc oxide, and titanium dioxide, have been reported to have intrinsic properties to kill a wide range of bacteria (Huang et al. 2017, Alves Claro et al. 2018). The ancient people in China, Egypt, and Greek used a silver knife and fork cutlery to avoid bacterial infection, and copper contained alloys were used for wound sterilization. The interaction between these antibacterial agents and bacteria is very complex and different possibilities to distract the biological processes of bacteria were used to explain the antibacterial activities. With the understanding that biofilms are difficult to eliminate from the surfaces, functional surfaces have been developed for long-lasting antibacterial activities through modification of the surface architectures (Wei et al. 2017, Pallavicini et al. 2017, Moreno-Couranjou et al. 2018, Paris et al. 2017, Ji et al. 2019). Different from those metal agents contained antibacterial surfaces, some flora and fauna developed self-cleaning surfaces through million years of evolution, such as lotus, cicada, and dragonfly, etc., using their superhydrophobic wettability.

Bioinspired artificial surfaces have been developed to prevent the adhesion of bacteria once they contact with the surfaces and prevent the formation of biofilms.

The discovery of penicillin by Alexander Fleming in 1928 was a revolution in fighting bacterial infections. Penicillin was produced by a fungus named *Penicillium*, which can inhibit the formation of the outside membrane of bacteria. After this, a series of antibiotic compounds have been produced which are derivates of penicillin to treat a wider range of bacterial infection. However, the effectiveness and easy access to these antibiotics have caused antibiotic resistance of some bacteria due to the overuse. Antibiotic resistance is an evolutionary response of bacteria through continued exposure to antibiotics, because any bacterium may have a gene resistance against the overused antibiotic, and the resistant bacteria can flourish when less nutritional competition in the community due to the removal of the sensitive bacteria. The WHO has classified antibiotic resistance as a serious threat all around the world. Thus, there has been strong global demands for advanced materials and surfaces for antibacterial applications.

BACTERIA

Bacteria are prokaryotes that consist of a single cell with some simple internal structures. Different criteria are used to classify bacteria, such as their cell walls, shape, or genetic makeup. The Gram staining method is always the first step for the identification of a bacteria which was developed by Hans Christian Gram in 1884. Gram staining procedures of bacteria are shown in Figure 2, in which bacteria are identified by the composition of the cell walls. Bacteria can be broadly categorized into Gram-positive and Gram-negative bacteria. Gram-positive bacteria have a thick peptidoglycan layer which can be stained by crystal violet (purple dye); Gram-negative bacteria have a thin layer of peptidoglycan which can be removed by alcohol and losing the stained color (Beveridge 2001).

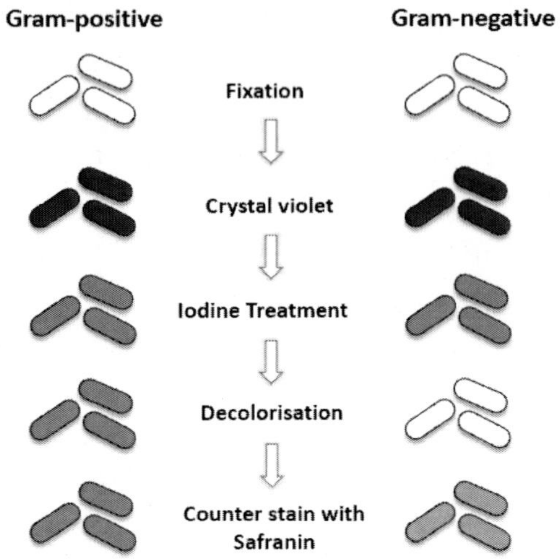

Figure 2. Procedure for the Gram staining and color change of Gram-positive and Gram-negative bacteria.

Schematic diagrams of the cell wall structures of the Gram-positive and Gram-negative bacteria are shown in Figure 3. The cell wall of the Gram-negative bacteria is composed of an outer membrane connected by lipoproteins to the thin layer of peptidoglycan (7-8 nm) in the periplasmic space between the outer and the inner lipid membranes. Porins in the outer membranes are channels to transport small hydrophilic molecules through the membrane, and lipopolysaccharide molecules are linked to the outer leaflet of the outer membrane. The cell wall structure of the Gram-positive bacteria also can be viewed in Figure 2, which is lack of an outer membrane, whilst a thick cell wall composed of a thick peptidoglycan layer (30-100 nm) is attached to lipoteichoic acids and teichoic acids, while lipoteichoic acids are extended to the inner cell membrane. Only one lipoprotein layer is located between the cell membrane and the peptidoglycan layer. It has shown that the cell wall of the Gram-negative bacteria is structurally and chemically more complex compare to the Gram-positive bacteria (Brown et al. 2015).

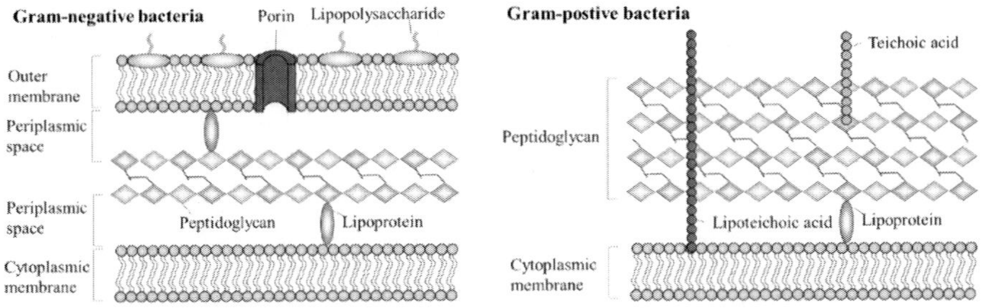

Figure 3. Schematic diagram of the cell wall structures of the Gram-negative and Gram-positive bacteria.

Bacteria have different shape and size, and cause different infections. Most of the Gram-negative bacteria have a rod-like shape around 2 μm long and 1 μm in diameter. Rod-like Gram-positive bacteria are longer than the rod-like Gram-negative bacteria. The only exception is the coccal like Gram-positive bacteria which is smaller with a diameter of about 0.6 μm. The size of bacteria can affect their interactions with the antibacterial surfaces.

The bacteria are typically cultured on the surface of solid nutrient media which normally consist of a mixture of protein digests, inorganic salts, and agar. Examples of the enriched media can support the growth of different bacteria include tryptic soy agar, chocolate agar, sheep blood agar. Bacteria grow on these solid media as colonies, and a colony is formed with a mass of bacteria all originating from a single mother cell, and hence the bacteria in the same colony are genetically alike.

There are four main growth phases of bacteria which are lag phase, log growth phase, stationary phase, and death phase. The lag phase is a period before the initiation of cell division. During this period, DNA, enzymes, and macromolecules are synthesized within the cell for reproduction, and cells may double or triple in size due to the preparation. When the cells are ready, bacteria start to divide at a maximum rate and then remain constant. This period is called the log phase because the logarithm of the number of bacteria increases linearly with time, and cells are most active during this period. The growth of the bacterial population will be ceased eventually during the stationary phase due to the exhaustion of available nutrients. The number of dying bacteria balance the number of newborn bacteria to stabilize the bacterial population. When the number of dying bacteria starts to exceed that of the newborn bacteria, viable cells start to decline exponentially during the death phase.

Antibacterial Strategies

The demands for antibacterial biomaterials are very broad, and a variety of approaches have been adopted to achieve antibacterial properties. The strategies to prevent bacterial infections can be classified into two major categories according to their action modes. As shown in Figure 4, the first strategy is antiadhesive surfaces, which can repel or prevent bacteria from adhering to surfaces by modifying the chemical or physical status of the surfaces. The second strategy is bactericidal surfaces, which can damage or kill the pathogenic bacteria once they contact the surfaces. A brief review of the well-established and newly developed antibacterial approaches are summarised in the following sections.

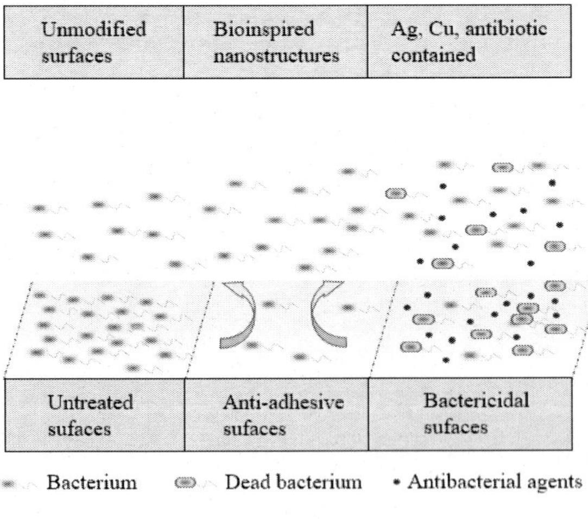

Figure 4. Schematic diagrams of different strategies currently used in the design of antibacterial surfaces.

Antiadhesive Surfaces

Bacterial adhesion is the first step for bacterial infection. It is clear that no possibility of bacteria colonization occurs if bacteria cannot adhere to a surface. Bacteria adhesion starts with contamination through dry states, wet conditions, or liquid carriers. The interactions between bacteria and the surfaces are critically influenced by various factors, such as surface morphology, environmental conditions, and bacteria features, etc. Direct airborne bacteria can be removed through adequate sterility procedures.

For those bacterial contamination transferred through aqueous solutions, which contain pathogen bacteria and proteins at the same time, it is known that some of the host proteins (collagen, laminin, fibrinogen, vitronectin, etc.) can promote bacterial adhesion and biofilm formation. Thus, these biomaterial surfaces are required to repel bacterial adhesion, whilst support the adsorption of these proteins for subsequent tissue integration. Bacterial adhesion in these situations is driven by surface hydrophobic interactions, hydrogen bonding, electrostatic functional groups. The bioinert surfaces can exert their antiadhesive function in protein-low or protein-rich conditions, depending on the surface morphologies or surface chemistry, such as self-assembled surfaces, polymer brush films, hydrogel films, and nanostructured surfaces. In some cases, the pre-adsorption process was applied to form a coating to change the interaction with bacteria, and it was reported that the adsorption of heparin can inhibit the adhesion of *S. epidermidis* by increasing of hydrophobicity. The electrostatic repulsion effect can be achieved by the polyanionic/polycationic functionalized surfaces because the Gram-negative bacteria have polyanionic glycocalyx and the Gram-positive bacteria have polycationic glycocalyx.

Some living organisms both animals and plants have developed some fascinating antiadhesive surfaces through million years of evolution to prevent colonization from the pathogens, such as the gecko skin, cicada wing, and dragonfly wing (Glinel et al. 2012). These superhydrophobic surfaces were obtained using the Cassie-Baxter wetting regime, which are low in surface energy and can reduce the contact between the bacteria and potential attach points. A bacteria contained droplet can keep the spherical shape and is easy to roll across the surface other than adhere to the surface. Besides, contaminants can adhere to the droplet rather than the antiadhesive surface and a self-cleaning activity is achieved, whilst this antifouling strategy is effective to both Gram-negative and Gram-positive bacteria.

Apart from the antiadhesive property of these superhydrophobic surfaces, it was also reported that these nanostructures also have an antibacterial effect through a physical damage approach. Usually, sharp nanorods with a diameter of 50-250 nm, and a height of 80-250 nm can penetrate or rupture the bacteria cell wall and cause leakage of nutrients, thereby leading to cell death. This contact killing mechanism has provided an effective approach to tackle the antibiotic-resistant bacteria (Cloutier, Mantovani, and Rosei 2015).

Recent studies have been carried out to investigate the relationship between the nanostructures, surface wettability, and antibacterial performance of the natural surfaces. Watson et al. reported that the hair-like structures of the gecko skin showed an antibacterial action to Gram-negative bacteria *P. gingivalis.* The superhydrophobic property of the skin inhibits the growth of the bacteria, and the stretching effect can cause damage to the cell wall (Watson et al. 2015). Ivanova et al. reported that the cicada wing has a surface full of nanoneedles with a 60 nm diameter tip and a height of 200 nm. The superhydrophobic nature of the surface has not fully prohibited the adhesion of the bacteria, but the nanostructured surface can kill the bacteria upon contact (Ivanova et al. 2012). Figure 5 shows the *P. aeruginosa* (Gram-negative) cells on the surface of the cicada wing. The cells were penetrated by the nanopillars thus leading to the death of the bacteria. This image verified that the nanostructured surface was effective to kill bacteria through the physical approach (Ivanova et al. 2012).

Jafar et al. pointed out that the killing-effect of the cicada wing is also related to the thickness difference between Gram-negative bacteria and Gram-positive bacteria. The peptidoglycan cell wall of Gram-positive bacteria is 4-5 times to that of the Gram-negative bacteria which makes it difficult to be penetrated. This implies that the rigidity of the cell walls significantly influences the survival ability of bacteria upon contact with the nanostructures (Hasan et al. 2013). Dragonfly wing has randomly distributed nanostructures on the surface, which was found to be efficient to kill both Gram-negative bacteria and Gram-positive bacteria (Ivanova et al. 2013). Kyle et al. found that the nanoscale topographies on the periodical cicada, annual DD cicada, and sanddragon dragonfly can rupture and kill adhered *S. cerevisiae*. They indicated that there is a connection between the geometry of the surface and cell death. High adhesion between cells and the surface led to a high degree of rupturing for a specific nanostructured surface (Nowlin et al. 2015). These findings from the natural antibacterial surfaces are useful for designing antimicrobial structures for large-scale cost-effective applications.

Figure 5. SEM image of the *Pseudomonas aeruginosa* cells on the surface of a cicada wing. The nanopillars are penetrated through the cells (Ivanova et al. 2012).

Bactericidal Surfaces

Inorganic antibacterial agents, such as silver, copper, zinc oxide, and titanium dioxide, have been reported to have intrinsic properties to kill a wide range of bacteria. Due to the complex interactions between the bacteria and the antimicrobial agents, there are different possibilities for these agents to distract the biological processes. It is difficult to give a general statement about the antibacterial mechanism of the bacteria-killing actions.

It was reported that silver ions can inactivate the thiol group in enzymes to kill the cells (Liau et al. 1997). Jeon *et al.* prepared the Ag-SiO_2 thin film through a sol-gel method which has an excellent antibacterial effect against *E. coli* and *S. aureus*. Ag ions were metalized and trapped in the SiO_2 matrix and showed a 99.9% killing rate after 24h contact (Jeon, Yi, and Oh 2003). Lee *et al.* tested the antibacterial performance of the Ag loaded coating on magnetic colloidal particles. It is thought that the release rate of the antibacterial Ag ions depended on the supply of the zerovalent Ag in the film (Lee, Cohen, and Rubner 2005). Reactive oxygen species were observed in some cases due to the silver ions (Liu, Wu, et al. 2010). Choi *et al.* reported that reactive oxygen species (including singlet oxygen, superoxide, hydrogen peroxide, and hydroxyl radical) generated by Ag can damage cellular constituents and affect cell functions, leading to the death of cells (Choi and Hu 2008).

Copper is reported to bind with some proteins to cause oxidative stress inside the cell and damage the surface. Besides, copper can degrade the integrity of the cell membrane, leading to leakage of nutrients and subsequent death of cells. Kuo *et al.* studied the effect of Cu content on the antibacterial effect of the Cu-Cr-N nanocomposite coating by DC reactive magnetron sputtering. Cu ions can be released when the sample was contacted with water which will damage enzymes on the cell surfaces (Kuo et al. 2007). Yao *et al.* prepared a Cu doped TiO_2 composite coating through micro-arc oxidation. The chemical state of Cu was investigated by XPS and Cu^{2+} was identified in the TiO_2 coating. The doping of Cu provided excellent antibacterial activities for the composite coating against both *E. coli* and *S. aureus*.

TiO_2 and ZnO can produce photocatalytic reactive oxygen species in the presence of UV light (Adams, Lyon, and Alvarez 2006, Cho et al. 2004). Adams *et al.* found that light was a significant factor to promote the generation of reactive oxygen species, although growth inhabitation was also supported in a dark environment (Adams, Lyon, and Alvarez 2006). TiO_2 is reputed with excellent antibacterial activity to both Gram-negative and Gram-positive bacteria. Rincon *et al.* indicated that the Gram-negative *E. coli* was more sensitive than the Gram-positive *B. subtilis* to the effect of TiO_2 (Rincon and Pulgarin 2005). However, Fu et al. found that TiO_2 has a higher killing rate to Gram-positive bacteria than Gram-negative bacteria (Fu, Vary, and Lin 2005). These results indicate that the sensitivity of bacteria to TiO_2 is determined by the specific interact conditions.

ZnO has been reported to be environmentally friendly material and is widely used as an additive ingredient due to its antibacterial property. Applerot et al. coated ZnO nanoparticles on the surface of the glass using an ultrasound irradiation method, which presented an excellent antibacterial effect against *E. coli* (Gram-negative) and *S. aureus* (Gram-positive). The antibacterial property of ZnO is attributed to the production of reactive oxygen species which can cause oxidative injury to the bacterial cell (Applerot, Lipovsky, et al. 2009, Applerot, Perkas, et al. 2009). Hu *et al.* deposited a zinc incorporated TiO_2 composite coating by plasma electrolytic oxidation method and discovered that the antibacterial activity of the Zn-incorporated coatings is enhanced compared with the Zn-free coatings, and the antibacterial effect is ascribed to the release of zinc ions (Hu et al. 2012).

Researchers have been inspired by these natural bactericidal surfaces that can kill bacteria coming in contact with them. Recently, some studies have been carried out to mimic these natural surfaces for developing the new generation of bio-inspired antibacterial surfaces (Cloutier, Mantovani, and Rosei 2015), and novel nanostructured surfaces were deposited to evaluate their antibacterial performances. The reactive ion etching method is a feasible way to fabricate nanostructured surface due to the etching effect. Hasan et al. employed the deep reactive ion etching to fabricate nanostructures on the Si surface, with pillars 220 nm in diameter. The surface was superhydrophobic with a contact angle of 154°. Antibacterial tests showed that 83% of *E. coli* and 86% of *S. aureus* were killed after 3 h contact with the surface (Ivanova et al. 2013).

Fisher et al. prepared a diamond nanocone surface to mimic the cicada fly wing. The nanodiamond films were fabricated by the microwave plasma chemical vapor deposition at about 850 °C which was modified by bias-assisted reactive ion etching to form the nanocone surfaces (Fisher et al.

2016). The results indicated that the nonuniform and low-density array exhibited a better antibacterial activity compared to the uniform and high-density array. The radius of the sharp tips of the nanocones ranged from 10 to 40 nm with a base width of 0.35 to 1.2 μm. It has verified that these tailor surfaces were effective to kill a broad range of microorganisms (Fisher et al. 2016).

Nanoneedles of black silicon surfaces have been fabricated by reactive ion etching which demonstrated excellent bactericidal activity against *P. aeruginosa*. May et al. used SEM and fluorescence microscopy to observe the killed bacteria, and the results indicated that the needle-like surface exhibited a higher antibacterial efficacy than the flat control surface (May et al. 2016).

Besides these nanostructure surfaces formed by reactive ion etching, it has been reported that sharp nanowires such as TiO_2 nanowires, ZnO nanowires, CNTs, and surfaces with other types of nanowires were also effective in killing microorganisms. Diu et al. prepared the TiO_2 nanowire surfaces with an alkaline hydrothermal method, and the diameter of these nanowires is around 100 nm. It has been found that these nanowires have enhanced the bactericidal effect against motile bacteria (*E. coli*) than that against the non-motile bacteria (*S. aureus*) due to more interactions between bacteria and the nanowires. Both brush and niche types of TiO_2 nanowire surfaces were verified to be effective to kill bacteria (Diu et al. 2014).

Chat et al. fabricated TiO_2 nanotubes on the surface of Ti foil using an anodization process. They found that the annealing temperature directly affected the antibacterial performance and photocatalytic dye degradation of the nanostructure surface. Gram-positive bacteria *B. atrophaeus* ATCC 9372 was applied for antibacterial tests. A higher antibacterial efficacy can be achieved from the combination of TiO_2 and UV illumination than from the UV illumination only (Chan et al. 2013).

Wang et al. prepared ZnO nanowire arrays with different orientations and used confocal laser scanning microscopy (CLSM) to investigate the antibacterial activity of the ZnO nanowires. The results showed that the antibacterial activity was strongly dependent on the ZnO nanoarray orientations with the randomly oriented ZnO nanoarrays exhibiting the superior antibacterial activity than the well-defined ZnO nanoarrays. The CLSM images showed that cell membranes were significantly damaged by the nanowires which can penetrate through the cells (Wang et al. 2007).

Densely packed ZnO nanowire arrays were synthesized using a solution-based method with a diameter of about 100 nm. The biological property of the densely packed ZnO nanowires was tested with the neuronal cell line, neonatal rat cardiomyocytes, and cardiac muscle cell line. The results revealed that the ZnO nanowire arrays had significant inhibitory effects against these cells in comparison to the control surfaces, including gold, glass, and polystyrene (Wang et al. 2017).

Kang et al. first provided direct evidence that SWCNTs have a strong antibacterial effect which can cause physical membrane damage, leading to the release of intracellular content. SEM images showed that the morphology (outer membrane) of *E. coli* changed after exposed to SWCNTs, while the control bacteria maintained their outer membrane structures (Kang et al. 2007). Besides, Kang et al. found that SWCNTs are more toxic against bacteria compare to that of MWCNTs. While bacteria expressed high levels of gene products after contacted with both MWCNT and SWCNT, yet compared with MWCNT, SWCNT can lead to a higher quantity and larger magnitude of gene expression (Kang et al. 2008).

Liu et al. reported that the individually dispersed CNTs were more toxic than the CNT aggregates against both Gram-negative and Gram-positive bacteria, such as *E. coli*, *P. aeruginosa*, *S. aureus*, and *B. subtilis*. The CNTs can be imagined as "nano darts" which can penetrate the cell walls of bacteria and lead the cell death. They inferred that the activity of the CNTs is related to the dispersion, concentration, and speed of the nano-darts (Liu et al. 2009). To study the interactions between single CNT and the bacteria membrane, Liu et al. carried out the piercing experiments on the cell walls using an AFM tip with a diameter of 2 nm. The results exhibited that no physical damage can be caused under the low load condition as shown in Figure 6, indicating that the antibacterial effect conferred by the single CNT is due to the accumulation of interactions between the CNT and the cell membrane (Liu, Ng, et al. 2010).

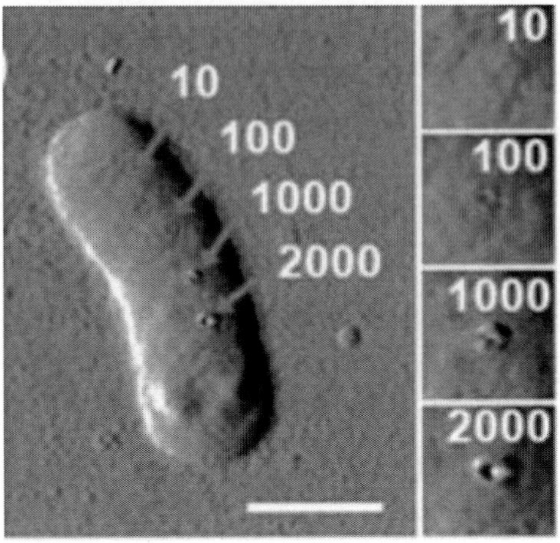

Figure 6. AFM image of *E. coli* after piercing by a 2 nm AFM tips under different load (10 nN – 2000 nN). Scale bar is 1μm (Liu, Ng, et al. 2010).

Conclusion and Future Perspectives

The control of bacterial infection is still a significant global healthcare problem even though vast efforts have been made to solve the problem through different approaches, such as novel functional biomaterials, surface coatings, surface architectures, etc. In this chapter, it is attempted to discuss the state-of-the-art antibacterial strategies to prevent the formation of biofilms. Antiadhesive surfaces, such as the sharp nanostructures on some insects and plants surface, which have shown effectiveness to repel the adhesion of bacteria. Bioinspired nanostructure surfaces have been produced using different etching procedures, which have the potential to be used in many biomedical tools. Further studies also indicated that these nanostructures also have a bactericidal effect by physically rupturing the cell membranes, but the long-term use toxicity especially interacting with human cells remains unknown. Some well-established techniques, such as physical vapor deposition, chemical vapor deposition, and plasma spray, are effective to fabricate bactericidal surfaces with modified antibacterial agents, or controllable release of these agents, which make it possible to deposit multi-functional surfaces combined antiadhesive and bactericidal properties are likely the antibacterial surfaces forward for preventing the formation of biofilms. It is envisaged that a better understanding of the antibacterial mechanism can support the design, deposition, and evaluation of novel antibacterial strategy in combating the challenge of bacterial infection and the problem of antibiotic resistance.

REFERENCES

Adams, L. K., D. Y. Lyon, and P. J. J. Alvarez. 2006. "Comparative eco-toxicity of nanoscale TiO_2, SiO_2, and ZnO water suspensions." *Water Research* 40 (19):3527-3532. doi: 10.1016/j.watres.2006.08.004.

Alves Claro, Ana Paula Rosifini, Reginaldo T. Konatu, Ana Lúcia do Amaral Escada, Miriam Celi de Souza Nunes, Cláudia Vianna Maurer-Morelli, Marcela Ferreira Dias-Netipanyj, Ketul C. Popat, and Diego Mantovani. 2018. "Incorporation of silver nanoparticles on Ti7.5Mo alloy surface containing TiO2 nanotubes arrays for promoting antibacterial coating – In vitro and in vivo study." *Applied Surface Science* 455:780-788. doi: https://doi.org/10.1016/j.apsusc.2018.05.189.

Applerot, G., A. Lipovsky, R. Dror, N. Perkas, Y. Nitzan, R. Lubart, and A. Gedanken. 2009. "Enhanced Antibacterial Activity of Nanocrystalline ZnO Due to Increased ROS-Mediated Cell Injury." *Advanced Functional Materials* 19 (6):842-852. doi: 10.1002/adfm.200801081.

Applerot, G., N. Perkas, G. Amirian, O. Girshevitz, and A. Gedanken. 2009. "Coating of glass with ZnO via ultrasonic irradiation and a study of its antibacterial properties." *Applied Surface Science* 256 (3):S3-S8. doi: 10.1016/j.apsusc.2009.04.198.

Beveridge, T. J. 2001. "Use of the Gram stain in microbiology." *Biotechnic & Histochemistry* 76 (3):111-118. doi: 10.1080/714028139.

Brown, L., J. M. Wolf, R. Prados-Rosales, and A. Casadevall. 2015. "Through the wall: extracellular vesicles in Gram-positive bacteria, mycobacteria and fungi." *Nature Reviews Microbiology* 13 (10):620-630. doi: 10.1038/nrmicro3480.

Chan, C. M. N., A. M. C. Ng, M. K. Fung, H. S. Cheng, M. Y. Guo, A. B. Djurisic, F. C. C. Leung, and W. K. Chan. 2013. "Antibacterial and photocatalytic activities of TiO_2 nanotubes." *Journal of Experimental Nanoscience* 8 (6):859-867. doi: 10.1080/17458080.2011.616540.

Cho, M., H. Chung, W. Choi, and J. Yoon. 2004. "Linear correlation between inactivation of E-coli and OH radical concentration in TiO_2 photocatalytic disinfection." *Water Research* 38 (4):1069-1077. doi: 10. 1016/j.watres.2003.10.029.

Choi, O., and Z. Q. Hu. 2008. "Size dependent and reactive oxygen species related nanosilver toxicity to nitrifying bacteria." *Environmental Science & Technology* 42 (12):4583-4588. doi: 10.1021/es703238h.

Cloutier, M., D. Mantovani, and F. Rosei. 2015. "Antibacterial Coatings: Challenges, Perspectives, and Opportunities." *Trends in Biotechnology* 33 (11):637-652. doi: 10.1016/j.tibtech.2015.09.002.

Diu, T., N. Faruqui, T. Sjostrom, B. Lamarre, H. F. Jenkinson, B. Su, and M. G. Ryadnov. 2014. "Cicada-inspired cell-instructive nanopatterned arrays." *Scientific Reports* 4. doi: ARTN 712210.1038/srep07122.

Fisher, L. E., Y. Yang, M. F. Yuen, W. J. Zhang, A. H. Nobbs, and B. Su. 2016. "Bactericidal activity of biomimetic diamond nanocone surfaces." *Biointerphases* 11 (1):1-6. doi: Artn 01101410.1116/1.4944062.

Fu, G. F., P. S. Vary, and C. T. Lin. 2005. "Anatase TiO_2 nanocomposites for antimicrobial coatings." *Journal of Physical Chemistry B* 109 (18):8889-8898.

Glinel, K., P. Thebault, V. Humblot, C. M. Pradier, and T. Jouenne. 2012. "Antibacterial surfaces developed from bio-inspired approaches." *Acta Biomaterialia* 8 (5):1670-1684. doi: 10.1016/j.actbio.2012.01.011.

Hasan, J., H. K. Webb, V. K. Truong, S. Pogodin, V. A. Baulin, G. S. Watson, J. A. Watson, R. J. Crawford, and E. P. Ivanova. 2013. "Selective bactericidal activity of nanopatterned superhydrophobic cicada Psaltoda claripennis wing surfaces." *Applied Microbiology and Biotechnology* 97 (20):9257-9262. doi: 10.1007/s00253-012-4628-5.

Hu, H., W. Zhang, Y. Qiao, X. Jiang, X. Liu, and C. Ding. 2012. "Antibacterial activity and increased bone marrow stem cell functions of Zn-incorporated TiO_2 coatings on titanium." *Acta Biomaterialia* 8 (2):904-915. doi: 10.1016/j.actbio.2011.09.031.

Huang, Pin, Kena Ma, Xinjie Cai, Dan Huang, Xu Yang, Jiabing Ran, Fushi Wang, and Tao Jiang. 2017. "Enhanced antibacterial activity and biocompatibility of zinc-incorporated organic-inorganic nanocomposite coatings via electrophoretic deposition." *Colloids and Surfaces B: Biointerfaces* 160:628-638. doi: https://doi.org/10.1016/j.colsurfb.2017.10.012.

Ivanova, E. P., J. Hasan, H. K. Webb, G. Gervinskas, S. Juodkazis, V. K. Truong, A. H. F. Wu, R. N. Lamb, V. A. Baulin, G. S. Watson, J. A. Watson, D. E. Mainwaring, and R. J. Crawford. 2013. "Bactericidal activity of black silicon." *Nature Communications* 4 (2838):1-7. doi: ARTN 283810.1038/ncomms3838.

Ivanova, E. P., J. Hasan, H. K. Webb, V. K. Truong, G. S. Watson, J. A. Watson, V. A. Baulin, S. Pogodin, J. Y. Wang, M. J. Tobin, C. Lobbe, and R. J. Crawford. 2012. "Natural Bactericidal Surfaces: Mechanical Rupture of Pseudomonas aeruginosa Cells by Cicada Wings." *Small* 8 (16):2489-2494. doi: 10.1002/smll.201200528.

Jeon, H. J., S. C. Yi, and S. G. Oh. 2003. "Preparation and antibacterial effects of Ag-SiO_2 thin films by sol-gel method." *Biomaterials* 24 (27):4921-4928. doi: 10.1016/S0142-9612(03)00415-0.

Ji, Xiaochao, Xiaoying Li, Yangchun Dong, Rachel Sammons, Linhai Tian, Helong Yu, Wei Zhang, and Hanshan Dong. 2019. "Synthesis and in-vitro antibacterial properties of a functionally graded Ag impregnated composite surface." *Materials Science and Engineering: C* 99:150-158. doi: https://doi.org/10.1016/j.msec.2019.01.087.

Kang, S., M. Herzberg, D. F. Rodrigues, and M. Elimelech. 2008. "Antibacterial effects of carbon nanotubes: Size does matter." *Langmuir* 24 (13):6409-6413. doi: 10.1021/la800951v.

Kang, S., M. Pinault, L. D. Pfefferle, and M. Elimelech. 2007. "Single-walled carbon nanotubes exhibit strong antimicrobial activity." *Langmuir* 23 (17):8670-8673.

Kuo, Y. C., J. W. Lee, C. J. Wang, and Y. J. Chang. 2007. "The effect of Cu content on the microstructures, mechanical and antibacterial properties of Cr-Cu-N nanocomposite coatings deposited by pulsed DC reactive magnetron sputtering." *Surface & Coatings Technology* 202 (4-7):854-860. doi: 10.1016/j.surfcoat.2007.05.062.

Lee, D., R. E. Cohen, and M. F. Rubner. 2005. "Antibacterial properties of Ag nanoparticle loaded multilayers and formation of magnetically directed antibacterial microparticles." *Langmuir* 21 (21):9651-9659. doi: 10.1021/la0513306.

Lewis, Kim, and Yue Shan. 2017. "Why tolerance invites resistance." *Science* 355 (6327):796.

Liau, S. Y., D. C. Read, W. J. Pugh, J. R. Furr, and A. D. Russell. 1997. "Interaction of silver nitrate with readily identifiable groups: relationship to the antibacterial action of silver ions." *Letters in Applied Microbiology* 25 (4):279-283. doi: DOI10.1046/j.1472-765X.1997.00219.x.

Lin, Ming-Hsien, Chi-Feng Hung, Ibrahim A. Aljuffali, Calvin T. Sung, Chi-Ting Huang, and Jia-You Fang. 2017. "Cationic amphiphile in phospholipid bilayer or oil–water interface of nanocarriers affects planktonic and biofilm bacteria killing." *Nanomedicine: Nanotechnology, Biology and Medicine* 13 (2):353-361. doi: https://doi.org/10.1016/j.nano.2016.08.011.

Liu, S. B., A. K. Ng, R. Xu, J. Wei, C. M. Tan, Y. H. Yang, and Y. A. Chen. 2010. "Antibacterial action of dispersed single-walled carbon nanotubes on Escherichia coli and Bacillus subtilis investigated by atomic force microscopy." *Nanoscale* 2 (12):2744-2750. doi: 10.1039/c0nr00441c.

Liu, S. B., L. Wei, L. Hao, N. Fang, M. W. Chang, R. Xu, Y. H. Yang, and Y. Chen. 2009. "Sharper and Faster "Nano Darts" Kill More Bacteria: A Study of Antibacterial Activity of Individually Dispersed Pristine Single-Walled Carbon Nanotube." *Acs Nano* 3 (12):3891-3902.

Liu, W., Y. A. Wu, C. Wang, H. C. Li, T. Wang, C. Y. Liao, L. Cui, Q. F. Zhou, B. Yan, and G. B. Jiang. 2010. "Impact of silver nanoparticles on human cells: Effect of particle size." *Nanotoxicology* 4 (3):319-330. doi: 10.3109/17435390.2010.483745.

May, P. W., M. Clegg, T. A. Silva, H. Zanin, O. Fatibello, V. Celorrio, D. J. Fermin, C. C. Welch, G. Hazell, L. Fisher, A. Nobbs, and B. Su. 2016.

"Diamond-coated 'black silicon' as a promising material for high-surface-area electrochemical electrodes and antibacterial surfaces." *Journal of Materials Chemistry B* 4 (34):5737-5746. doi: 10.1039/c6tb01774f.

Moreno-Couranjou, Maryline, Rodolphe Mauchauffé, Sébastien Bonot, Christophe Detrembleur, and Patrick Choquet. 2018. "Anti-biofouling and antibacterial surfaces via a multicomponent coating deposited from an up-scalable atmospheric-pressure plasma-assisted CVD process." *Journal of Materials Chemistry B* 6 (4):614-623. doi: 10.1039/C7TB02473H.

Nowlin, Kyle, Adam Boseman, Alan Covell, and Dennis LaJeunesse. 2015. "Adhesion-dependent rupturing of Saccharomyces cerevisiae on biological antimicrobial nanostructured surfaces." *Journal of the Royal Society Interface* 12.

Pallavicini, Piersandro, Barbara Bassi, Giuseppe Chirico, Maddalena Collini, Giacomo Dacarro, Emiliano Fratini, Pietro Grisoli, Maddalena Patrini, Laura Sironi, Angelo Taglietti, Marcel Moritz, Ioritz Sorzabal-Bellido, Arturo Susarrey-Arce, Edward Latter, Alison J. Beckett, Ian A. Prior, Rasmita Raval, and Yuri A. Diaz Fernandez. 2017. "Modular approach for bimodal antibacterial surfaces combining photo-switchable activity and sustained biocidal release." *Scientific Reports* 7 (1):5259. doi: 10.1038/s41598-017-05693-3.

Paris, Jean-Baptiste, Damien Seyer, Thierry Jouenne, and Pascal Thébault. 2017. "Elaboration of antibacterial plastic surfaces by a combination of antiadhesive and biocidal coatings of natural products." *Colloids and Surfaces B: Biointerfaces* 156:186-193. doi: https://doi.org/10.1016/j.colsurfb.2017.05.025.

Rincon, A. G., and C. Pulgarin. 2005. "Use of coaxial photocatalytic reactor (CAPHORE) in the TiO_2 photo-assisted treatment of mixed E-coli and Bacillus sp and bacterial community present in wastewater." *Catalysis Today* 101 (3-4):331-344.

Swartjes, J. J. T. M., P. K. Sharma, T. G. van Kooten, H. C. van der Mei, M. Mahmoudi, H. J. Busscher, and E. T. J. Rochford. 2015. "Current Developments in Antimicrobial Surface Coatings for Biomedical

Applications." *Current Medicinal Chemistry* 22 (18):2116-2129. doi: 10.2174/0929867321666140916121355.

Tripathy, A., P. Sen, B. Su, and W. H. Briscoe. 2017. "Natural and bioinspired nanostructured bactericidal surfaces." *Advances in Colloid and Interface Science* 248:85-104. doi: 10.1016/j.cis.2017.07.030.

Wang, X. L., F. Yang, W. Yang, and X. R. Yang. 2007. "A study on the antibacterial activity of one-dimensional ZnO nanowire arrays: effects of the orientation and plane surface." *Chemical Communications* (42):4419-4421. doi: 10.1039/b708662h.

Wang, Y. C., Y. Wu, F. Quadri, J. D. Prox, and L. Guo. 2017. "Cytotoxicity of ZnO Nanowire Arrays on Excitable Cells." *Nanomaterials* 7 (4). doi: ARTN 8010.3390/nano7040080.

Watson, G. S., D. W. Green, L. Schwarzkopf, X. Li, B. W. Cribb, S. Myhra, and J. A. Watson. 2015. "A gecko skin micro/nano structure - A low adhesion, superhydrophobic, anti-wetting, self-cleaning, biocompatible, antibacterial surface." *Acta Biomaterialia* 21:109-122. doi: 10.1016/j.actbio.2015.03.007.

Wei, Ting, Zengchao Tang, Qian Yu, and Hong Chen. 2017. "Smart Antibacterial Surfaces with Switchable Bacteria-Killing and Bacteria-Releasing Capabilities." *ACS Applied Materials & Interfaces* 9 (43):37511-37523. doi: 10.1021/acsami.7b13565.

In: An Introduction to Antibacterial Properties ISBN: 978-1-53618-305-4
Editor: Nicholas Paquette

Chapter 2

How It Works: Major Mechanism of Antibacterial Agents

Hee Su Kim and Dong Gun Lee*
School of Life Sciences, College of Natural Sciences,
Kyungpook National University, Daegu, Republic of Korea

Abstract

Antibacterial agent is a natural or synthetic compound which inhibits the growth of bacteria in microorganisms. These agents determine their antibacterial range according to the type of targeted bacteria and are commonly divided into gram-positive bacteria and gram-negative bacteria. If agents act on both gram-positive and -negative bacteria, it can be classified as a broad-spectrum antibacterial agent. Moreover, if only a few of them work on it, they can be classified as a medium-spectrum antibacterial agent. In general, antibacterial agents are categorized based on structural features and mode of action. These include five main topics: inhibition of cell wall synthesis, disturbance of nucleic acid synthesis, protein synthesis, folic acid and cell membrane dysfunction by changing cell membrane permeability. This chapter focused on characteristic

*Corresponding Author's Email: dglee222@knu.ac.kr.

mechanism of five main categories and the corresponding representative type of antibacterial agents.

Keywords: antibacterial agent, gram-positive bacteria, gram-negative bacteria, cell membrane dysfunction

INTRODUCTION

Antibacterial is an agent that interferes with the growth and reproduction of bacteria. These agents kill or prevent bacteria by fighting against bacterial. Heat, chemicals such as chlorine, and all antibiotic drugs have antibacterial properties [1-3]. Although the history of antibiotics has not been very long, hundreds of them have been developed to date, including synthetic antibiotics, as well as natural ones [4-6]. The discovery of these antibacterial agents has greatly changed the aspect of human life. Recently, however, the emergence of antibiotic-resistant bacteria that are resistant to all of antibiotics has become a serious social problem [7, 8]. Thus, the development of new antibiotics and the market for antibiotics continues to grow [9].

Most antibiotics could be classified differently according to chemical structure, mechanisms and antibacterial spectrum [10, 11]. First, different antibacterial agents have variant antibacterial spectrum. Antibacterial agents that work on both gram-positive and gram-negative bacteria are called broad spectrum antibiotics, and antibacterial agents that could only kill or suppress some microorganisms are classified as narrow spectrum antibiotics [12-14].

Moreover, antibacterials are now most commonly described as agents used to disinfect surfaces and eliminate potentially harmful bacteria. Antibacterials according to their speed of action and residue production are divided into two groups [15]. Bacteriostats, disinfectants, sanitizers and sterilizers are different groups of antibacterials. As such, bacteriostats inhibit the growth and reproduction of bacteria and allow the body's immune system to work together. On the other hand, bactericidal agents kill the bacteria that are growing, but they cannot kill the bacteria that are in a stationary phase that do not multiply [16]. Thus, antibacterials are classified based on Inhibition of cell wall synthesis, Inhibition of protein synthesis, Inhibition of bacterial nucleic acid synthesis [17-19].

Based on the preceding content, we will examine the characteristics of the five main mechanism sof antibacterial agents and learn more about the types of antibacterial agents.

INHIBITION OF CELL WALL SYNTHESIS

Bacteria are surrounded by a structure called a cell wall that is not present in human cells, so they can maintain much higher intra-bacterial pressure than osmotic pressure in humans [20]. Bacteria are destroyed when synthesis is suppressed at each stage of the synthesis of cell wall, which function as essential for the survival of bacteria. Antibiotics that inhibits the synthesis of bacterial cell wall include a β-lactams (penicillin, cephalosporin, etc.) and vancomycin. Then, cell wall synthesis inhibitors usually have antibacterial effects only on bacteria that are proliferating [19, 21-23]. However, some bacteria, such as mycoplasma without cell wall, have no antibacterial effect no matter how much antibiotics are administered that inhibits cell wall synthesis, and do not affect human cells because they do not have cell wall.

β-Lactams

β-lactam antibiotic has a pharmacophore, β-lactam ring, and is a sterilized antibiotic that deactivates enzymes by binding them to active part of PBP (penicillin-binding protein), a multifunctional enzyme involved in cross-linkage of bacterial cell wall, the peptidoglycan chain [24, 25]. Because PBP is an enzyme unique to bacteria, β-lactam is a representative antibiotic with selective toxicity, but β-lactam antibiotic that fails to penetrate the outer membrane of the gram-negative bacteria does not have an antibacterial effect on gram-negative bacteria. β-lactam antibiotic is divided into penicillin, cephalosporin, carbapenem and monobactam, which have only β-lactam ring, depending on the ring-shaped structure that form the β-lactam ring [26, 27]. There is also an oxopenam that could impede the β-lactam decomposition enzyme with little antibacterial activity. A widely used group of antibiotics in clinic, most of which use semi-synthetic antibiotic.

Immediately after its first use, penicillin had a very dramatic effect on the treatment of the *Staphylococcus* infection, but as it use increased, *Staphylococcus* produced β-lactamase and began to show resistance to penicillin [28]. β-lactamase hydrolysis the β-lactam ring to eliminate antibacterial effect. The strains that produce β-lactamase gradually increased, and eventually penicillin became useless for treatment of *Staphylococcus*. In the 1960s, methicillin, which is not hydrolyzed by the β-lactamase, was developed, including oxacillin and nafcillin. In 1961, ampicillin was developed, and it had antibacterial effect because it penetrates the outer membrane of some gram-negative bacilli and binds to their PBP [29]. Since then, carboxypenicillin has excellent antibacterial effect against gram-negative bacilli, showing antibacterial effect even to *Pseudomonas aeruginosa* [30].

Cephalosporins

Cephalosporins are divided into 1st to 4th generation according to antibacterial spectrum and characteristics and currently the most commonly used in clinical [31]. Cephalosporins vary depending on the type of causative bacteria, not the more severe the disease is, the higher generation of antibiotics they choose [27, 32]. 1st generation cephalosporins are antibacterial to the *Staphylococcus* that produce the β-lactamase because the β-lactam ring is not hydrolyzed by the β-lactamase produced in the *Staphylococcus*. In addition, they have good antibacterial effect in other gram-positive bacteria except for *Enterococcus* and Methicillin-resistant *Staphylococcus aureus* (MRSA). However, other gram-negative bacteria and anaerobic bacteria, including *Haemophilus influenzae*, do not have antibacterial effect. The 2nd generation cephalosporins have a wider antibacterial spectrum for gram-negative bacteria compared to the 1st generation, and their antibacterial effect have also improved for anaerobic bacteria. 3rd generation cephalosporins have better antibacterial effect against gram-negative than 1st and 2nd generation and could be largely divided into those with good or no antibacterial effect against *Pseudomonas aeruginosa*. Some of the drugs that are good for *Pseudomonas aeruginosa* are Ceftazidime and Cefoperazone, and those that have weak antibacterial effect in the *Pseudomonas aeruginosa* but have good antibacterial effect in the gram-positive are Cefotaxime, Ceftizoxime and Ceftriaxone [33]. The characteristics of 4th generation cephalosporins are that the stability of β-lactamase is better than before, and they have the advantage of rapidly penetrating the outer membrane of gram-negative bacteria [34].

Monobactams

Monobactam is a drug composed exclusively of β-lactam ring, which has very good antibacterial effect against gram-negative bacteria, whereas gram-positive or anaerobic bacteria have very little antibacterial effect [35, 36]. Also, it could also be used for patients who have hypersensitivity to β-lactam antibiotics.

Glycopeptides

Glycopeptide antibiotic is a powerful antibiotic that works by inhibiting the synthesis of bacterial cell walls and has a relatively narrow antibacterial spectrum that mainly acts on gram-positive bacteria [21, 37]. This includes vancomycin and Teicoplanin [38, 39]. Vancomycin is obtained from *Streptomyces orientalis*. In the late 1970s, the use of vancomycin increased as the MRSA gradually increased. Over the years, vancomycin maintained excellent antibacterial effect for almost all gram-positive bacteria, but the incidence of patients by Vancomycin-resistant *Enterococcus* (VRE) has increased significantly until recently since the first report of VRE in Europe in 1986 [7, 40, 41]. Teicoplanin is an antibiotic derived from *Actinoplanes teichomyceticus* and the side effect of vancomycin, red-man syndrome, is not very visible.

Disturbance of Protein Synthesis

For bacteria to proliferate, the proteins needed for proliferation must be synthesized within the cytoplasm. If this protein synthesis is inhibited, bacteria will not be able to increase. Aminoglycoside, tetracycline, macrolide, lincosamide and chloramphenicol, etc. indicate antibacterial activity by inhibiting the protein synthesis of bacteria [42, 43].

Aminoglycoside

Antibiotics produced in *Streptomyces* and *Micromomospora* are complicated in structure and are mostly made through biosynthesis. Aminoglycoside travels through a protein passage called porin in the outer membrane of gram-negative bacteria to the space around the cytoplasm and then into the cell by active transport [44, 45]. Aminoglycoside is irreversible combined with bacterial 30S ribosome to inhibit protein synthesis and indicate antibacterial activity [46]. It is characterized by good antibacterial effect in most gram-negative bacteria, but anaerobic bacteria do not have antibacterial effect and do not pass through Blood-Brain Barrier (BBB). For example, streptomycin binds to S12 protein of 30S ribosome to inhibit initiation of translation by suppressing the binding of formylmethionyl-tRNA, which is initiation tRNA [47].

Macrolide

Macrolide, which means a large ring, combines with bacterial 50S ribosome to inhibit protein synthesis [48-50]. It has a tetrakaidecagon, pentadecagon and hexadecagon ring structure and includes erythromycin, azithromycin and clarithromycin. These are good for bacteria such as *Mycoplasma*, *Chlamydia*, *Legionella* and *Campylobacter*. It also has similar antibacterial effect in gram-positive bacteria, *Neisseria*, *Haemophilus influenzae* and anaerobic bacteria [51]. In addition, macrolide antibiotics could be administered instead of penicillin to patients who overreact to penicillin because they also have a good antibacterial effect on penicillin-sensitive bacteria.

Erythromycin combines reversible with bacterial 50S ribosome to pause peptide bonds and exhibit antibacterial effect while inhibiting protein synthesis [52, 53]. In particular, gram-positive bacteria are distributed at concentrations about 100 times higher in the cytoplasm than gram-negative bacteria, indicating stronger antibacterial effect.

Clarithromycin has half antibacterial effect against *Haemophilus influenzae*, but it has the same or more twice effect for most of the anaerobic bacteria [54, 55]. In addition, it is more useful in treating *Helicobacter pylori* or *Mycobacterium* infections [56].

Tetracycline

When chlortetracycline extracted from *Streptomyces aureofaciens* was first introduced in 1948, it became known as a wide range of antibiotics as it showed excellent antibacterial effect on various kinds of bacteria such as aerobic gram-positive bacteria, gram-negative bacteria, anaerobic bacteria, *Rickettsia*, *Mycoplasma* and *Chlamydia* [57]. Since the confirmation of tetracycline-resistant bacteria in 1953, resistance to various bacteria has increased, making it difficult to use tetracycline [58-60]. However, recently developed tigecycline is a new type of glycylcycline antibiotics that has antibacterial effect on resistant-bacteria such as MRSA, VRE and Drug-resistant *Streptococcus pneumoniae* [61].

Lincomycin

Lincomycin is separated from *Streptomyces lincolnensis* and antibacterial effect is similar to erythromycin but has different chemical properties [62]. It works well for most anaerobic bacteria.

Chloramphenicol

Chloramphenicol is a substance found in *Streptomyces venezulae* that has antibacterial effect while inhibiting the function of protein synthesis in bacteria [53, 63]. It is a broad-spectrum antibiotic that not only inhibit the gram-negative and gram-positive bacteria but also sterilize various microorganisms such as *Rickettsia*, *Mycoplasma* and *Chlamydia*. In particular, chloramphenicol was widely used to prevent and treat diseases of livestock because it had excellent antibacterial effect on various microorganisms without harming livestock. However, it has been banned from being used in humans and livestock since it was reported to have bone marrow toxicity [64].

Disturbance of Nucleic Acid Synthesis

Antibiotics that inhibit nucleic acid interfere with the translation of DNA and transcription of RNA, a process necessary for bacterial proliferation. In the process of cloning a bacteria DNA, an enzyme celled DNA gyrase is involved to prevent the entanglement of each strand of double helix [65-67]. Quinolone antibiotics combine with the A subunit of this enzyme to inhibit bacterial growth [68]. In addition, the antituberculous agent, rifampin combines with DNA-dependent RNA polymerase to interfere with RNA synthesis.

Quinolone

Quinolone is a synthetic compound that inhibits the enzyme called DNA gyrase, preventing DNA replication [69]. Initially, quinolone antibiotics were only used for urinary tract infections because they were limited to some gram-negative bacteria and had many side effects. However, in 1980, by attaching a fluorine to the structure of these drugs, we expanded the range from gram-negative bacteria to gram-positive bacteria [70]. In addition, by attaching a piperazine, they had an antibacterial effect on the *Pseudomonas aeruginosa*, and compared to the quinolone of the order generation, they improved absorption rate and reduced the side effects [71]. Quinolones are different in structure from original antibiotics, so it is difficult to express resistant bacteria, but the number of resistant bacteria is increasing rapidly [72].

Disturbance of Folic Acid Synthesis

The drugs related to synthetic inhibition of folic acid, an important precursor to nucleic acid synthesis, are being used as an antibiotic. Folic acid is not biosynthesis in humans, but bacteria use folic acid by biosynthesis within themselves. Therefore, drugs that impede the process of folic acid synthesis could interfere with bacteria without affecting the human cells.

Sulfonamide is a typical anti-folate and inhibits activity of dihydropteroic acid synthase (DHPS) as an analogue of ρ-aminobenzoic acid (PABA), which is one of the substrates of DHPS [73]. PABA is needed in enzymatic reactions that produce folic acid, which acts as a coenzyme in the synthesis of purines and pyrimidines [74]. Inhibiting the synthesis of tetrahydrofolate (THF) results in reduced DNA synthesis, which stops bacterial growth.

CELL MEMBRANE DYSFUNCTION

The cell membrane controls the components inside the cell by performing active transport as a permeability barrier. This change in permeability causes polymers or ions to escape the cells and kill the cells [75]. Bacterial membranes are different from those of human cells, so selective chemotherapy is possible. In other words, antibiotics acting on cell membranes changes the permeability of cell membranes, causing bacteria to lose balance between inside and outside of the membrane, leading to cell death [21]. But mass administration could also cause toxicity to human cells. Some drugs that inhibit cell membrane include polymyxin, colistin, gramicidin and daptomycin.

Polymyxin

Polymyxins are cationic, surface-active agents that disrupt the structure of cell membrane phospholipids and increase cell permeability by a detergent-like action [76]. Gram-negative bacteria are much more sensitive than Gram-positive bacteria because they contain more phospholipid in their cytoplasm and outer membranes. Acquired resistance is rare but can occur with Pseudomonas aeruginosa as a result of decreased bacterial permeability. Polymyxins are highly active against many Gram-negative bacteria, including Pseudomonas aeruginosa but not Proteus. Activity against P. aeruginosa is reduced *in vivo* by calcium at physiological concentrations.

Gramicidins

Gramicidin is active against most gram-positive bacteria and against select gram-negative organisms. The soil bacterium *Bacillus brevis* produces a mixture of short polypeptides, gramicidins A, B, and C, which are collectively called gramicidin an important antibiotic that effectively kills gram-positive bacteria [77]. Analysis of the mechanism of action of gramicidin showed that it causes a loss of ions from the bacteria against which it is effective. Gramicidin acts as an ionophore, i.e., a substance that one can add to a lipid bilayer and thereby increase greatly the rate at which ions move across it [78]. Thus, gramicidin is clearly a channel and indeed has served as an excellent model for the behavior of membrane channels [79].

CONCLUSION

We divide antibacterial mechanisms into five categories, and the most representative classification groups are cell wall synthesis inhibition, cell membrane dysfunction, protein synthesis inhibition, nucleic acid synthesis inhibition, and folic acid synthesis inhibition. Based on what was described earlier, antibacterial agents can be largely divided into bacteriostat and bactericidal. Suppressing bacterial cell wall synthesis and losing the function of the cell membrane is a bactericidal because it kills proliferating bacteria. On the other hand, inhibiting the synthesis of bacteria's proteins, nucleic acids and folic acids is a bacteriostat because they do not directly kill the bacteria that multiply but cause bacteria to stop reproducing. In addition to the previously known drugs, various antibiotics such as trimethdprim, bacitracin, mupirocin and oxazolidione have been developed and used. Due to this diversity of antibiotics, prescriptions are now required considering the pharmacokinetic characteristics, antibacterial spectrums, and antibacterial mechanisms of each antibacterial agent.

REFERENCES

[1] Dodd, MC; et al., Interactions of fluoroquinolone antibacterial agents with aqueous chlorine: reaction kinetics, mechanisms, and

transformation pathways. *Environmental science & technology*, 2005, 39(18), p. 7065-7076.

[2] Eddy, RS; et al., An *in vitro* evaluation of the antibacterial efficacy of chlorine dioxide on E. faecalis in bovine incisors. *Journal of Endodontics*, 2005, 31(9), p. 672-675.

[3] Conesa, C; et al., Effect of heat treatment on the antibacterial activity of bovine lactoferrin against three foodborne pathogens. *International journal of dairy technology*, 2010, 63(2), p. 209-215.

[4] Butler, MS; Buss, AD. Natural products—the future scaffolds for novel antibiotics? *Biochemical pharmacology*, 2006, 71(7), p. 919-929.

[5] Silver, LL. Are natural products still the best source for antibacterial discovery? The bacterial entry factor. *Expert opinion on drug discovery*, 2008, 3(5), p. 487-500.

[6] Ramón-García, S; et al., Targeting Mycobacterium tuberculosis and other microbial pathogens using improved synthetic antibacterial peptides. *Antimicrobial agents and chemotherapy*, 2013, 57(5), p. 2295-2303.

[7] Ahmed, MO; Baptiste, KE. Vancomycin-resistant enterococci: a review of antimicrobial resistance mechanisms and perspectives of human and animal health. *Microbial Drug Resistance*, 2018, 24(5), p. 590-606.

[8] Furuya, EY; Lowy, FD. Antimicrobial-resistant bacteria in the community setting. *Nature Reviews Microbiology*, 2006, 4(1), p. 36-45.

[9] Moellering Jr, RC. Discovering new antimicrobial agents. *International journal of antimicrobial agents*, 2011, 37(1), p. 2-9.

[10] Etebu, E; Arikekpar, I. Antibiotics: Classification and mechanisms of action with emphasis on molecular perspectives. *Int. J. Appl. Microbiol. Biotechnol. Res*, 2016, 4(2016), p. 90-101.

[11] Peach, KC; et al., Mechanism of action-based classification of antibiotics using high-content bacterial image analysis. *Molecular BioSystems*, 2013, 9(7), p. 1837-1848.

[12] Acar, J. Broad-and narrow-spectrum antibiotics: an unhelpful categorization. *Clinical Microbiology and Infection*, 1997, 3(4), p. 395-396.
[13] van Saene, R; Fairclough, S; Petros, A. Broad-and narrow-spectrum antibiotics: a different approach. *Clinical Microbiology and Infection*, 1998, 4(1), p. 56-57.
[14] Williams, DJ; et al., Narrow vs broad-spectrum antimicrobial therapy for children hospitalized with pneumonia. *Pediatrics*, 2013, 132(5), p. e1141-e1148.
[15] Nemeth, J; Oesch, G; Kuster, SP. Bacteriostatic versus bactericidal antibiotics for patients with serious bacterial infections: systematic review and meta-analysis. *Journal of Antimicrobial Chemotherapy*, 2015, 70(2), p. 382-395.
[16] Heizmann, P; Heizmann, WR. *Bacteriostatic-bactericidal.* Medizinische Klinik (Munich, Germany: 1983), 2007, 102(9), p. 720-726.
[17] Abbanat, D; Macielag, M; Bush, K. Novel antibacterial agents for the treatment of serious Gram-positive infections. *Expert opinion on investigational drugs*, 2003, 12(3), p. 379-399.
[18] Yong, AL; et al., Investigation of antibacterial mechanism and identification of bacterial protein targets mediated by antibacterial medicinal plant extracts. *Food chemistry*, 2015, 186, p. 32-36.
[19] Chung, YC; et al., Relationship between antibacterial activity of chitosan and surface characteristics of cell wall. *Acta pharmacologica sinica*, 2004, 25(7), p. 932-936.
[20] Dmitriev, B; Toukach, F; Ehlers, S. Towards a comprehensive view of the bacterial cell wall. *Trends in microbiology*, 2005, 13(12), p. 569-574.
[21] Bush, K. Antimicrobial agents targeting bacterial cell walls and cell membranes. *Rev Sci Tech*, 2012, 31(1), p. 43-56.
[22] Bugg, TD; et al., Bacterial cell wall assembly: still an attractive antibacterial target. *Trends in biotechnology*, 2011, 29(4), p. 167-173.
[23] Green, DW. The bacterial cell wall as a source of antibacterial targets. *Expert opinion on therapeutic targets*, 2002, 6(1), p. 1-20.

[24] Cho, H; Uehara, T; Bernhardt, TG. Beta-lactam antibiotics induce a lethal malfunctioning of the bacterial cell wall synthesis machinery. *Cell*, 2014, 159(6), p. 1300-1311.

[25] Georgopapadakou, N. Penicillin-binding proteins and bacterial resistance to beta-lactams. *Antimicrobial agents and chemotherapy*, 1993, 37(10), p. 2045.

[26] Majiduddin, FK; Materon, IC; Palzkill, TG. Molecular analysis of beta-lactamase structure and function. *International journal of medical microbiology*, 2002, 292(2), p. 127.

[27] Petri, W. *Penicillins, cephalosporins, and other β-lactam antibiotics*. Goodman and Gilman's The Pharmacological Basis of Therapeutics 12th Ed McGraw-Hill, New York, 2011, p. 1477-1504.

[28] Adam, D. Global antibiotic resistance in Streptococcus pneumoniae. *Journal of Antimicrobial Chemotherapy*, 2002. 50(suppl_1), p. 1-5.

[29] Bartzatt, R; Cirillo, S; Cirillo, JD. Molecular properties and antibacterial activity of the methyl and ethyl ester derivatives of ampicillin. *Physiological chemistry and physics and medical NMR*, 2004, 36(2), p. 85-94.

[30] Gupta, NN; Lakhe, MM. Brief overview of antibacterial agents. *The Journal of the Association of Physicians of India*, 2010, 58, p. 8-12.

[31] Flynn, EH. *Cephalosporins and penicillins: chemistry and biology*. 2013, Elsevier.

[32] Marshall, WF; Blair, JE. The cephalosporins. in *Mayo Clinic Proceedings*. 1999, Elsevier.

[33] Angelescu, M; Apostol, A. Cefepime (maxipime), large spectrum 4th generation cephalosporin, resistant to beta-lactamases. *Chirurgia* (Bucharest, Romania: 1990), 2001, 96(6), p. 547-552.

[34] Giamarellou, H. Fourth generation cephalosporins in the antimicrobial chemotherapy of surgical infections. *Journal of chemotherapy*, 1999, 11(6), p. 486-493.

[35] Singh, GS. β-Lactams in the new millennium. Part-I: monobactams and carbapenems. *Mini reviews in medicinal chemistry*, 2004, 4(1), p. 69-92.

[36] Asbel, LE; Levison, ME. Cephalosporins, carbapenems, and monobactams. *Infectious Disease Clinics,* 2000, 14(2), p. 435-447.
[37] Van Bambeke, F; et al., Glycopeptide antibiotics. *Drugs*, 2004, 64(9), p. 913-936.
[38] Nagarajan, R. Structure-activity relationships of vancomycin-type glycopeptide antibiotics. *The Journal of antibiotics*, 1993, 46(8), p. 1181-1195.
[39] Nagarajan, R. Antibacterial activities and modes of action of vancomycin and related glycopeptides. *Antimicrobial Agents and chemotherapy*, 1991, 35(4), p. 605.
[40] French, G. Enterococci and vancomycin resistance. *Clinical Infectious Diseases*, 1998, 27(Supplement_1), p. S75-S83.
[41] Leclercq, R; Courvalin, P. Resistance to glycopeptides in enterococci. *Clinical Infectious Diseases*, 1997, 24(4), p. 545-554.
[42] Cocito, C; et al., Inhibition of protein synthesis by streptogramins and related antibiotics. *The Journal of antimicrobial chemotherapy*, 1997, 39(suppl_1), p. 7-13.
[43] Sutcliffe, JA. Antibiotics in development targeting protein synthesis. *Annals of the New York Academy of Sciences*, 2011, 1241(1), p. 122-152.
[44] Hancock, R; et al., Interaction of aminoglycosides with the outer membranes and purified lipopolysaccharide and OmpF porin of Escherichia coli. *Antimicrobial agents and chemotherapy*, 1991, 35(7), p. 1309-1314.
[45] Garneau-Tsodikova, S; Labby, KJ. Mechanisms of resistance to aminoglycoside antibiotics: overview and perspectives. *Med Chem Comm*, 2016, 7(1), p. 11-27.
[46] Mehta, R; Champney, WS. 30S ribosomal subunit assembly is a target for inhibition by aminoglycosides in Escherichia coli. *Antimicrobial agents and chemotherapy*, 2002, 46(5), p. 1546-1549.
[47] Carr, JF; Gregory, ST; Dahlberg, AE. Severity of the streptomycin resistance and streptomycin dependence phenotypes of ribosomal protein S12 of Thermus thermophilus depends on the identity of highly

conserved amino acid residues. *Journal of bacteriology*, 2005, 187(10), p. 3548-3550.

[48] Poehlsgaard, J; Douthwaite, S. Macrolide antibiotic interaction and resistance on the bacterial ribosome. *Current Opinion in Investigational Drugs*, 2003, 4(2), p. 140-148.

[49] Hansen, JL; et al., The structures of four macrolide antibiotics bound to the large ribosomal subunit. *Molecular cell*, 2002, 10(1), p. 117-128.

[50] Gaynor, M; Mankin, AS. Macrolide antibiotics: binding site, mechanism of action, resistance. *Current topics in medicinal chemistry*, 2003, 3(9), p. 949-960.

[51] Mabe, S; Eller, J; Champney, WS. Structure–Activity Relationships for Three Macrolide Antibiotics in Haemophilus influenzae. *Current microbiology*, 2004, 49(4), p. 248-254.

[52] Usary, J; Champney, WS. Erythromycin inhibition of 50S ribosomal subunit formation in Escherichia coli cells. *Molecular microbiology*, 2001, 40(4), p. 951-962.

[53] Siibak, T; et al., Erythromycin-and chloramphenicol-induced ribosomal assembly defects are secondary effects of protein synthesis inhibition. *Antimicrobial agents and chemotherapy*, 2009, 53(2), p. 563-571.

[54] Miyazaki, S; et al., Efficacy of azithromycin, clarithromycin and β-lactam agents against experimentally induced bronchopneumonia caused by Haemophilus influenzae in mice. *Journal of Antimicrobial Chemotherapy*, 2001, 48(3), p. 425-430.

[55] Thadepalli, H; et al., Comparison of telithromycin, a new ketolide, with erythromycin and clarithromycin for the treatment of Haemophilus influenzae pneumonia in suckling, middle aged and senescent mice. *International journal of antimicrobial agents*, 2002, 20(3), p. 180-185.

[56] Leung, WK; Graham, DY. Clarithromycin for Helicobacter pylori infection. *Expert opinion on pharmacotherapy*, 2000, 1(3), p. 507-514.

[57] Pereira, JF; et al., Extraction of tetracycline from fermentation broth using aqueous two-phase systems composed of polyethylene glycol

and cholinium-based salts. *Process Biochemistry*, 2013, 48(4), p. 716-722.

[58] Trzcinski, K; et al., Expression of resistance to tetracyclines in strains of methicillin-resistant Staphylococcus aureus. *Journal of Antimicrobial Chemotherapy*, 2000, 45(6), p. 763-770.

[59] Schmitt, H; et al., Tetracyclines and tetracycline resistance in agricultural soils: microcosm and field studies. *Microbial ecology*, 2006, 51(3), p. 267-276.

[60] Michalova, E; Novotna, P; Schlegelova, J. Tetracyclines in veterinary medicine and bacterial resistance to them. A review. *Veterinarni Medicina-UZPI (Czech Republic)*, 2004.

[61] Zhanel, GG; et al., The glycylcyclines. *Drugs*, 2004, 64(1), p. 63-88.

[62] Spížek, J; Řezanka, T. Lincomycin, cultivation of producing strains and biosynthesis. *Applied microbiology and biotechnology*, 2004, 63(5), p. 510-519.

[63] Dowling, PM. Chloramphenicol, thiamphenicol, and florfenicol. *Antimicrobial therapy in veterinary medicine*, 2013, p. 269-277.

[64] Ambekar, C; et al., Metabolism of chloramphenicol succinate in human bone marrow. *European journal of clinical pharmacology*, 2000, 56(5), p. 405-409.

[65] Alt, S; et al., Inhibition of DNA gyrase and DNA topoisomerase IV of Staphylococcus aureus and Escherichia coli by aminocoumarin antibiotics. *Journal of antimicrobial chemotherapy*, 2011, 66(9), p. 2061-2069.

[66] Heide, L. New aminocoumarin antibiotics as gyrase inhibitors. *International Journal of Medical Microbiology*, 2014, 304(1), p. 31-36.

[67] Ehmann, DE; Lahiri, SD. Novel compounds targeting bacterial DNA topoisomerase/DNA gyrase. *Current opinion in pharmacology*, 2014, 18, p. 76-83.

[68] Drlica, K; et al., Quinolone-mediated bacterial death. *Antimicrobial agents and chemotherapy*, 2008, 52(2), p. 385-392.

[69] Hooper, DC; Jacoby, GA. Topoisomerase inhibitors: fluoroquinolone mechanisms of action and resistance. *Cold Spring Harbor perspectives in medicine*, 2016, 6(9), p. a025320.

[70] Mitsuhashi, S; et al., Fluorinated quinolones—new quinolone antimicrobials, in *Progress in Drug Research/Fortschritte der Arzneimittelforschung/Progrès des recherches pharmaceutiques*. 1992, Springer. p. 9-147.

[71] Foroumadi, A; et al., Synthesis and antibacterial activity of new fluoroquinolones containing a substituted N-(phenethyl) piperazine moiety. *Bioorganic & medicinal chemistry letters*, 2006, 16(13), p. 3499-3503.

[72] Blondeau, JM. Fluoroquinolones: mechanism of action, classification, and development of resistance. *Survey of ophthalmology*, 2004, 49(2), p. S73-S78.

[73] Sköld, O. Sulfonamide resistance: mechanisms and trends. *Drug resistance updates*, 2000, 3(3), p. 155-160.

[74] Whiteley, R. *Effect of multiple-antibiotic treatments on the evolution of antibiotic resistance in Pseudomonas aeruginosa*. 2014, University of Oxford.

[75] Kohanski, MA; et al., Mistranslation of membrane proteins and two-component system activation trigger antibiotic-mediated cell death. *Cell*, 2008, 135(4), p. 679-690.

[76] Bradshaw, JP. Cationic antimicrobial peptides. *BioDrugs*, 2003, 17(4), p. 233-240.

[77] Haggag, WM. Isolation of bioactive antibiotic peptides from Bacillus brevis and Bacillus polymyxa against Botrytis grey mould in strawberry. *Archives of Phytopathology and Plant Protection*, 2008, 41(7), p. 477-491.

[78] Takada, Y; Matsuo, K; Kataoka, T. Gramicidin A directly inhibits mammalian Na+/K+-ATPase. *Molecular and cellular biochemistry*, 2008, 319(1-2), p. 99-103.

[79] Andersen, OS; Koeppe, RE; Roux, B. Gramicidin channels. *IEEE transactions on nanobioscience*, 2005, 4(1), p. 10-20.

In: An Introduction to Antibacterial Properties ISBN: 978-1-53618-305-4
Editor: Nicholas Paquette

Chapter 3

ANTIMICROBIAL PROPERTIES OF (METH)ACRYLATE BASED HYDROGELS

Simonida Lj. Tomić and Jovana S. Vuković
Faculty of Technology and Metallurgy, University of Belgrade, Belgrade, Serbia

ABSTRACT

The rapid emergence of antibiotic-resistant pathogens is becoming an imminent global public health problem. The development of novel antimicrobial materials aiming to prevent or control infections caused by these pathogens is a very important issue. Polymeric hydrogels are versatile materials, which can be a great alternative to conventional treatments of infections. Because of its high hydrophilicity, unique three-dimensional network, fine biocompatibility, and cell adhesion, the hydrogels are suitable biomaterials for drug delivery in antimicrobial areas. The biocompatible nature of hydrogels makes them a convenient starting platform to develop selectively active antimicrobial materials. Hydrogels with antimicrobial properties can be designed by loading of known antimicrobial agents, or the material itself can be designed to possess inherent antimicrobial activity. The combination of polymeric hydrogels based on (meth)acrylate with metal ions (Ag, Cu, and Zn) is a simple and effective approach for obtaining multicomponent systems with diverse functionalities. Silver (Ag^{+}), copper (Cu^{2+}), and zinc (Zn^{2+}) ions have been loaded into hydrogels for antimicrobial applications. The incorporation of metal ions into hydrogels not only enhances the antimicrobial activity of hydrogels but also influences their swelling and release characteristics. Herein, we summarize recent findings in (meth)acrylate hydrogels containing metal ions with antimicrobial properties.

Keywords: hydrogels, 2-hydroxyethyl (meth)acrylate, poly(N-vinylpyrrolidone), itaconic acid, metal ions, swelling, controlled release, antimicrobial properties, biomedical applications

INTRODUCTION

The infectious diseases caused by pathogenic microorganisms such as bacteria, viruses, and parasites are still a public health problem despite the major development in health care and medical technology. Infection remains one of the most serious complications for human health and in some severe cases can lead to death. Most acute sequelae and global mortality were caused predominantly by infectious diseases (Global Burden of Disease Study 2017). For this reason, great efforts have been made to find new antibiotics with low toxicity to the host cell and strong activity against a wide range of bacterial pathogens that do not generate bacterial resistance. However, the antimicrobial agents and peptides themselves usually suffer from environmental toxicity, short-term antimicrobial activity, and proteolytic instability and degradation. To overcome such disadvantages, antimicrobial agents are often physically incorporated or chemically conjugated with biocompatible polymers, such as hydrogels, to increase their antimicrobial efficacy and specificity, reduce their cytotoxicity, extend their biostability and biocompatibility, and finally promote other biomimetic physicochemical properties.

Hydrogels with antimicrobial properties are of considerable biomedical interest and can be obtained using different strategies that can be summarized in two major methodologies. First, hydrogels can be used as carriers for antimicrobial agents. These agents can be different nature, such as nanoparticles (usually gold or silver), antibiotics or antimicrobials. Second, hydrogels can be constructed using materials which exhibit intrinsic antimicrobial activity (Malmsten 2011). The most relevant exemplars of inherently antimicrobial materials used to produce hydrogels include antimicrobial peptides, chitosan, or synthetic polymers with antimicrobial functional groups.

Since the first discovery of penicillin (Tan and Tatsumura 2015), antibiotics have been widely used in the antibacterial field. With the evolution of public hygiene and biomedical technology, many infections have been effectively suppressed or even overcome, and the quality of life of human beings has been significantly improved. However, a serious issue that remains is that the use of antibiotics has led to the emergence of multidrug-resistant microorganisms, which are very difficult to combat (Boehle et al. 2017). This has led to over 13 million people dying per year from infectious diseases worldwide (Song and Jang 2014). The most concerning fact were that the corresponding antibiotic-resistant bacteria emerged almost immediately after the advanced antibiotics were approved, e.g., the fidaxomicin-resistant *Enterococci* (K-1476) and the methicillin-resistant *Staphylococcus aureus* (*S. aureus*) (MRSA) (Zipperer et al. 2016; Molton et al. 2013; van Hoek et al. 2011). Synthetic antibacterial agents, such as salicylate, chlorhexidine, isothiazolinone, thiosemicarbazone, octenidine, and quaternary ammonium compounds, also face constant threats because of the drug resistance acquired by microorganisms (Malmsten 2014). Additionally, the application of conventional antibiotics brings other issues, such as solubility, overdose, and cytotoxicity. Therefore, there is a high demand for an effective and safe drug delivery system, which can reduce the risk of bacterial drug-resistance and regulate the toxicity of antibacterial drugs.

As challenges to the growing threats from pathogenic microorganisms resistant to drugs, researchers are studying a variety of advanced antibacterial materials. Among them, heavy metal ions and natural extracts were applied in the antibacterial treatment. However, these materials can inhibit and kill not only pathogenic microbes but also normal cells in the human body, which limits their potential application (García-Barrasa 2010).

Despite the tremendous ability of antimicrobial hydrogels in breaking down multidrug-resistant microbes, the interactions between antimicrobial polymers and microbial cell membranes are nonspecific, which in most cases, cause mammalian cell death above certain concentrations (Ng et al. 2014). One solution is to combine antibiotics and antimicrobial hydrogels so that the less antimicrobial hydrogel is used and the associated toxicity is minimized.

Another type of interesting hydrogels are those that contain antimicrobial metal nanoparticles. The incorporation of silver ions and silver nanoparticles in hydrogels initiated substantial advances in wound treatment (Ng et al. 2014). The silver nanoparticles supported by PVA/cellulose acetate/gelatin were successfully prepared. The hydrogels have antimicrobial activity against various fungi and bacteria (Abd El-Mohdy 2013). The toxicity of silver and other metal salts is a disadvantage for this type of hydrogels. Efforts have been made to reduce the toxicity.

Hydrogels are a form of 3D porous materials, which consist of polymeric chains with either physical or chemical crosslinking (Liu 2018; Zheng et al. 2017; Wei 2017; Caliari and Burdick 2016). Hydrogels have been extensively studied as alternative materials for antibacterial applications. By carefully selecting monomers and crosslinkers, the desired abilities of hydrogels, such as the hydrophilicity and porosity, can be developed for antibacterial applications. Moreover, some types of hydrogels also have an inherent antibacterial property. According to the classification of hydrogel matrices and the antibacterial agents, the antibacterial hydrogels can be divided into three types: (i) inorganic nanoparticles containing hydrogels, (ii) antibacterial agent containing hydrogels, and (iii) hydrogels with inherent antibacterial capabilities.

Hydrogels with inherent antibacterial activity refer to polymers which exhibit antimicrobial activity by themselves or those whose biocidal activity is conferred through their chemical modification, not including hydrogels with incorporated antimicrobial organic compounds or active inorganic systems (Malmsten 2011). These hydrogels developed in recent years can be classified as novel antimicrobial materials without traditional defects.

Antibacterial polymers include nonstimulated antibacterial polymers and potential antibacterial polymers. The most common nonstimulated antibacterial polymers have certain components in their structures that are important for antibacterial action. The hydrogels composed of thermoresponsive PNIPAM and redox-responsive polyferrocenylsilane macromolecules exhibited strong antibacterial activities while maintaining high biocompatibility (Sui et al. 2013). The redox-induced formation of hydrogel-Ag composites showed good antimicrobial activity against *E. coli*. pH-sensitive and thermal-sensitive hydrogels based on HEMA and IA copolymers possess potential for biomedical applications, especially for skin treatments and wound dressings (Tomić et al. 2010). P(HEMA/IA) could block the entry of *S. aureus* and *E. coli* into hydrogel dressing. Also, no evidence of cell toxicity or considerable hemolytic activity was observed in an *in vitro* study of P(HEMA/IA) biocompatibility. Hydrogels prepared by photopolymerization of PEG diacrylate and monomer containing ammonium salt (RNH_3Cl) demonstrated both antibacterial and antifouling properties (La et al. 2011). The potential antibacterial polymers present a class of polymers that could be converted to become antibacterial under certain conditions, such as exposure to light. The photodynamic poly(2-hydroxyethyl methacrylate-*co*-methyl methacrylate) (P(HEMA-*co*-MAA)) copolymers crosslinked by porphyrin were reported to be promising for the prevention of intraocular lens-associated infectious endophthalmitis (Parsons et al. 2009). Another photodynamic PHEMA-based hydrogel also exhibited a light-induced bactericidal effect through the release of nitric oxide (Halpenny et al. 2009). These antibacterial polymers provided not only antibacterial materials but also responsive delivery and release procedures.

Acrylate based polymers are currently used in many fields of biomedicine such as corneal prosthesis, intraocular lenses and contact lenses in ophthalmology (Chirila and Harkin 2016), bone cement for orthopedic applications (Shalaby et al. 2007), tissue engineering (Van Blitterswijk and De Boer 2014), due to their excellent properties such as biocompatibility and suitable mechanical performance, among others (Stoy and Climent 1996). Many of these acrylate products for various applications have been approved by the US Food and Drug Administration (FDA) and are expected to produce massively. However, many of their potential biomedical uses are sometimes hindered by their low mechanical strength, biological interactions, electrical and/or thermal properties, fluid sorption and transport, antibacterial activity, porosity, when they are synthesized as scaffolds for tissue engineering applications. Thus, new advanced acrylate based materials have been developed and are currently under intensive research to solve all these problems using multicomponent polymeric systems or by combination with other materials and/or nanomaterials to form composites or nanocomposites with or without interconnected porous morphology.

Fluid sorption and transport are also very important in biomedicine because they play a very important role in cell survival, especially in tissue engineering applications (Van Blitterswijk and De Boer 2014). Thus, acrylate hydrogels such as poly(2-hydroxyethyl methacrylate) or poly(2-hydroxyethyl acrylate), are very important hydrophilic materials as these polymers were able to absorb and swell retaining large amounts of fluid within their structure (Serrano-Aroca et al. 2004; Clayton et al. 1997; Monleón-Pradas et al. 2001; Monleón-Pradas et al. 2001). The excellent water sorption property has made these types of materials very promising in a wide range of biomedical applications such as controlled drug delivery, tissue engineering, wound healing (Stoy and Climent 1996; Ahmed 2015).

The ability of hydrogels to absorb water arises from hydrophilic functional groups attached to the polymeric backbone, while their resistance to dissolution arises from crosslinks between network chains (Tanaka 1981). However, these single network hydrogels have weak mechanical properties and slow response at swelling. Therefore, they need reinforcement, as already mentioned, which can also modify their water sorption properties. For example, the combination of hydrophilic and hydrophobic functional groups of acrylate polymers as multicomponent polymeric systems.

There are many acrylate hydrogels, which exhibit a non-Fickian diffusion behavior such as poly(2-hydroxyethyl acrylate) (Monleón-Pradas et al. 2001; Monleón-Pradas et al. 2001). Studies have shown that the swelling of poly(2-hydroxyethyl methacrylate) is controlled by Fickian diffusion (Gehrke et al. 1995). Thus, copolymeric hydrogels based on 2-hydroxyethyl methacrylate (HEMA) and epoxy methacrylate (EMA) synthesized by bulk polymerizations showed that the swelling process of these polymers also follows Fickian behavior and the equilibrium water content (EWC) decreased with increase in EMA content due to its hydrophobicity (Wang and Wu 2005). Remarkably, the pH has a big influence on the swelling properties and diffusion mechanism of acrylate based materials. Thus, the swelling capacity of polymeric networks was decreased in acidic pH, while a reverse trend was seen in basic pH. However, these hydrogels were found to be hydrolytically stable in phosphate buffer, which makes them potential materials for biomedical applications (Monleón-Pradas et al. 2001).

In biomedicine, bacterial infections can lead to implant failure, which may cause major economic losses and suffer among patients, despite the use of preoperative antibiotic prophylaxis and the aseptic processing of materials. Therefore, novel antibacterial materials are urgently needed (Shi et al. 2016). However, acrylates themself do not have antibacterial activity intrinsically, thus some comonomeric components, fillers, and antibacterial agents need to be incorporated (He et al. 2012).

In the field of dental materials, since methyl methacrylate was firstly used in tooth restoration in 1937, biocompatible and bioadhesive methacrylate monomers have been extensively used as dental materials (Hi et al. 2012). The most commonly used methacrylate monomers in formulations of commercial dental resin-based materials are methyl methacrylate (MMA), 2,2-bis[4-(2-hydroxy-3-methacryloyloxypropyl)-phenyl]propane (Bis-GMA),1,6-bis-[2-methacryloyl oxyethoxycarbonylamino]-2,4,4-trimethyl hexane (UDMA) and tri-ethyleneglycol dimethacrylate (TEGDMA) (Moszner and Salz 2001). However, the dental materials produced using these monomers are not antibacterial, which is very important in this biomedical field. Thus, another strategy to design acrylate hydrogels with desired antibacterial performance implicates the incorporation of silver nanoparticles (Ag NPs). This modification produced a strong antibacterial activity against *Escherichia coli* and also improved the mechanical properties of acrylate resins for dental applications. Such antibacterial effects were mainly attributed to the release of silver ions upon immersion of the dental composite in water, which appeared to be fairly non-toxic to humans (Guo et al. 2013). Poly(methyl methacrylate) (PMMA) nanofibers containing silver nanoparticles were synthesized by radical-mediated dispersion polymerization and also showed enhanced antimicrobial efficacy, compared to silver sulfadiazine and silver nitrate at the same silver concentration (Kong and Jang 2008).

Infections are frequent and highly undesired occurrences after orthopedic procedures. Besides, the growing concern caused by the rise in antibiotic resistance progressively decreased the efficacy of such drugs. The integration of silver nanoparticles in the polymeric mineralized acrylate based nanocomposites provides antibacterial activity against bacteria (González-Sánchez et al. 2015). Thus, a series of hydrogels were synthesized by crosslinking of Ag/graphene composites with acrylic acid and N, N′-methylene bisacrylamide at different mass ratios. In this study, prepared hydrogel with an optimal Ag to graphene mass ratio of 5:1 exhibited much stronger antibacterial abilities than other hydrogels and showed excellent biocompatibility, high swelling ratio, and good extensibility at the same time. *In vivo* experiments indicated that these nanocomposite hydrogels could significantly accelerate the healing rate of artificial wounds in rats, successfully reconstructing intact and thickened epidermis during 15 days of healing of impaired wounds (Fan et al. 2014). In the same way, acrylic acid (AA) grafted onto poly(ethylene terephthalate) (PET) film through gamma ray-induced graft copolymerization with silver nanoparticles on the surface showed strong and stable antibacterial activity (Ping et al. 2011).

Hydrogels as polymeric biomaterials have been examined for 50 years because of their properties and potential applications in medicine and pharmacy (Peppas et al., 2000; Okay et al. 2009; Singh et al. 2014; Jiao et al. 2006). Hydrogels are cross-linked polymeric materials that can swell in water and biological fluids but do not dissolve in them. Due to their high water content, they have a soft consistency and a favorable degree of flexibility that are very similar to living tissues, resulting in favorable biocompatibility. In the swollen state, they can reach a mass of 10% to tens of thousands more compared to the dry gel-xerogel mass. Water retention capacity and permeability are very important properties of hydrogels. Crosslinking between different polymeric chains results in a viscoelastic, sometimes totally elastic behavior and contributes to the gel having adequate hardness and elasticity. The choice of initial components and fraction in the synthesis provides the ability to adjust the hydrogel structure, mechanical, thermal, morphological properties, as well as swelling and external stimuli sensitivity (Gulrez et al. 2011; Himi and Mauria 2013; Hoffman 2012; Chai et al. 2017; Pal et al. 2009; Das 2013;).

A group of hydrogels showing changes in swelling, in response to the action of stimuli from the environment, such as pH, temperature, ionic strength of the solution, intensity of light, magnetic and electric field strength, belong to intelligent materials. Due to its specific properties, the application of hydrogels in the field of medicine and pharmacy has increasing importance and scope. Hydrogels must remain mechanically strong enough during the swelling, i.e., to retain their geometry and shape, to be flexible, and to be able to release the loaded active agent molecules at a controlled rate. For certain bioengineering applications, the properties of mimicry are important, i.e., imitating the properties of living tissues.

Biocompatibility is one of the most significant characteristics of newly synthesized hydrogels to be used in medicine and pharmacy. It implies the interaction between the material and the organism in which the material is applied. To find use as medical implants and devices, they must satisfy certain criteria. An important feature of the material is that it must possess harmlessness, which means that its application does not cause cytotoxicity. They should not cause unwanted reactions to the body, irritation, or allergies, such as mutagenicity and carcinogenicity. If the tissue cannot accept the foreign body, then it is unacceptable for making the implants. In a preliminary biocompatibility assessment, *in vitro* cytotoxicity tests are most often used. If the newly synthesized materials are not biocompatible, i.e., if they show a cytotoxic effect on normal cells, they could induce necrosis (accidental cell death) or apoptosis (programmed cell death) and thus could not be used for human use. This is the first phase in the research and development of new potential pharmaceutical and medical products.

Most of the toxicity problems of hydrogels are related to the unreacted components during synthesis. Therefore, the understanding of the toxicity of various monomers used as constituents of hydrogels is crucial. The relationship between chemical structures and the cytotoxicity of acrylate and methacrylate monomers has been studied. Some of the measures taken to solve this problem include modifying the polymerization kinetics to achieve higher conversion, as well as the rinsing of the hydrogel. The formation of hydrogels without the initiator supplements was examined to eliminate the problem of a lagging initiator. The most commonly used technique is gamma radiation (Ristić et al. 2011).

Infection is the most common complication associated with the use of medical devices (catheters, prostheses, surgical sutures, implants). One of the most well-known polymers, which in itself possesses to some extent antimicrobial properties is poly(vinyl pyrrolidone) (PVP). One of the most important testing of polymers for biomedical applications is the evaluation of their antimicrobial activity. The most common causes of infection are known, and these microorganisms are used to test the antimicrobial potential of new materials (Rodriguez-Hernandez 2016; Salome and Schneider 2013).

Poly(vinyl pyrrolidone) (PVP) is a frequently used polymer for the synthesis of semi-interpenetrating hydrogel networks (Wei et al. 2011; Bajpai et al. 2008; Wang et al. 2011). Hydrogels based on PVP may be candidates for various biomedical applications, such as controlled drug release systems (Mishra et al. 2008; Erizal et al. 2013). PVP possesses good biocompatibility and can be used as the main component in the synthesis of materials used for temporary skin dressings and bandages (Erizal et al. 2013; Ajji et al. 2005; De Silva 2011; Biazar et al. 2012; Sohail et al. 2014). Its use as biomaterials in artificial blood plasma dominated the Second World War. Lately, polymers based on PVP have been tested quite a lot. PVP is a linear, water-soluble, synthetic polymer that is widely used in medicine, such as a blood plasma extender, a carrier for drug delivery systems, tissue regeneration, and implant replacement (Barros et al. 2011). Because of its special molecular structure, PVP has many outstanding properties. PVP has satisfied biocompatibility and hydrophilic properties, which have been used for composite tissue engineering matrices. It is one of the most frequently used interpenetrating polymer because it can be expected to influence on morphological, swelling, and drug release properties of hydrogels (Abdelrazek et al. 2013; Chadha et al. 2006; Domingues_et al. 2013; Erizal et al. 2013; Giri et al. 2011; Marsano et al. 2005; Naghdeali and Adimi 2015; Tomar and Sharma 2013; Wang and Wang 2010; Wei et al. 2014; Yanpeng et al. 2006).

(Meth)acrylates (2-hydroxyethyl methacrylate (HEMA) and 2-hydroxyethyl acrylate (HEA)) are commonly used monomeric components for the syntheses of hydrogels, due to its biocompatibility and high hydrophilicity. HE(M)A based hydrogels found a lot of applications in pharmaceutical and biomedical fields. It can be used for biomaterials as coatings, intraocular lenses, tissue scaffolding biomaterials, and controlled release drug delivery systems. To improve mechanical stability or swelling properties of hydrogels based on HE(M)A, they can be combined with hydrophilic polymers as blends or semi-IPNs (Prasitsilp et al. 2003; Barrett et al. 1986; Menapace et al. 1989; Tomić et al. 2006; Babić et al. 2015; Baino 2010; Sanna et al. 2012; Chen et al. 2007).

Itaconic acid (IA) is a monomeric component that can contribute to the specific properties of hydrogels. Itaconic acid can be produced from renewable sources by fermentation. IA is highly hydrophilic and is expected to show favorable biocompatibility due to its natural origin. Small amounts of IA enhances a favorable pH sensitivity and thus the degree of swelling of the hydrogel can be adjusted (Bera et al. 2015; Gils et al. 2011; Okabe et al. 2009; Petruccioli et al. 1999; Rashid et al. 2016; Sakthivel et al. 2014; Sariri and Jafarian 2002; Sudarkodi et al. 2012).

Heavy metals, such as silver, gold, copper, zinc, have been used to fight against microbes for a long time. The advantage of metal ions incorporated in hydrogels over other forms is in the possibility to control the release of the ions, manage the dosage required for the therapeutic effect and the patient's exposure, owing to the unique characteristics of the hydrogels. Also, the resistance to these antimicrobial agents was found rare. Silver is recognized and proved as an extraordinary antimicrobial agent for centuries, from ancient to modern civilization. Even before people knew about microorganisms and infections, they found that everyday use of dishes and cutlery made from silver allowed longer preservation of food and beverage. Further, ancient Egyptians applied silver powder to heal wounds.

Silver shows strong antimicrobial activity against a broad spectrum of Gram-positive, Gram-negative bacteria, and yeasts (Percival et al. 2011; Hossain et al. 2019). Among them are some of the most frequent causes of healthcare-associated infections and hospital outbreaks, with developed multidrug resistance, such as *S. aureus*, *P. aeruginosa,* and *C. albicans* (Singh and Singh 2012; Percival et al. 2011; Lipsky and Hoey 2009). Consequently, various forms of silver found many biomedical applications, including the healing of both acute and chronic wounds (Jaiswal and Koul 2016; Bhowmick and Koul 2016). Certainly, the most common use of silver is in the form of nanoparticles (Ag NPs). According to recent studies, the primary mechanism of the antibacterial action of Ag NPs is the release of Ag^{+} (Pelgrift and Friedman 2013; Guo et al. 2013; Le Ouay and Stellacci 2015). Despite the widespread use of silver, bacterial resistance to Ag^{+} has been found rare and developed slowly, owing to the multiple possible mechanisms of antimicrobial action of silver, unlike antibiotics with a single mechanism of action in most cases (Lara et al. 2010).

Copper, as an oligoelement and micronutrient, is present in certain quantities in the human body, actively participating in numerous biochemical processes. For the normal functioning of the body, a continuous optimum intake of this nutrient is required. Owing to the great antimicrobial potential, nowadays copper is widely employed as an antimicrobial agent. The copper-made surfaces were recommended in conditions where a high level of hygiene is required, because of the excellent activity against the most common pathogens *S. aureus*, *P. aeruginosa,* and *E. coli* (Grass et al. 2011; Mikolay et al. 2010; Gould et al. 2009). To prevent the incidence and further development of infection on stainless steel surfaces, which are often in use in clinical conditions, households, schools, and kindergartens, copper nanoparticles loaded poly (ethylene glycol) diacrylate hydrogels have been synthesized as antibacterial coatings (Cometa et al. 2013). Copper nanoparticles (Cu NPs) and copper oxide nanoparticles (CuO NPs) exhibited significant microbiocidal activities against both fungi (*Saccharomyces cerevisiae, Candida* species) and bacteria (*E. coli, S. aureus,* and *Klebsiella pneumoniae*) (Kruk et al. 2015). Although there is not much literature about their mechanisms of action, it is assumed that copper interacts with amino and carboxyl groups on the surfaces of the microbial cells, which induces the formation of reactive oxygen species (ROS), inhibiting DNA replication and synthesis of amino acids in microbial cells (Pelgrift and Friedman 2013).

Copper is an essential element in the process of wound healing. It promotes the growth of the blood vessels, by inducing the formation of the vascular endothelial growth factor (VEGF) (Xie and Kang 2009). Also, copper has a stimulating effect on the proliferation of epithelial cells, fibroblasts, and collagen (Hu 1998; Simeon et al. 2000; Harris et al. 1980). Cu^{2+} ions loaded alginate hydrogels provide effective healing by controlling inflammatory processes and wound infections (Klinkajon and Supaphol 2014). Chitosan and copper-based nanocomposites found to accelerate wound healing, activating cells, cytokines, and growth factors at different stages of the healing process (Gopal et al. 2014).

Zinc is also one of the oligoelements and as a cofactor of numerous enzymes is involved in biochemical processes in the human body (Lansdown et al. 2007). It is considered essential for processes such as collagen synthesis, antibody production, as well as cell proliferation, which are necessary for the normal course of wound healing (Ashfaq et al. 2016). As one of the antimicrobial agents, zinc exhibits antibacterial activity against *S. aureus*, *E. coli*, *Streptococcus pyogenes, Listeria monocytogenes* and *Salmonella* (Xu et al. 2015; Jones et al. 2008; Liu et al. 2009). Zinc is frequently studied and applied as an antimicrobial agent in the form of oxide nanoparticles, ZnO NPs. It is considered that the antimicrobial activity of ZnO NPs is initiated by the presence of soluble Zn^{2+} ions. Some of the general mechanisms involved in the antibacterial activity are cell membrane blebbing and disruption. The release of Zn^{2+} ions results in the generation of reactive oxygen species (ROS), which further causes DNA disruption and breakage, protein oxidation and lipid peroxidation (Agarwal et al. 2018;

Huh and Kwon 2011). It was found that zinc alginate hydrogels prepared by the internal setting method, as potential wound dressings, exhibited significant antibacterial activity against *E. coli* (Straccia et al. 2015). To develop novel materials for wound healing application, the hydrogel films fabricated comprising sodium carboxymethylcellulose and hydroxypropylmethylcellulose in the presence of citric acid (CA) and zinc oxide nanoparticles (ZnO NPs) by solution casting method, showed great antibacterial potential against *S. aureus* (Dharmalingam et al. 2020).

This review presents interesting, "intelligent" polymeric systems created to have antimicrobial properties. Monomeric components (HEMA/HEA and itaconic acid (IA)), poly(vinyl pyrrolidone) (PVP) polymer as well as active metal ions (silver, copper and zinc) are selected to design potent antimicrobial polymeric systems. Their swelling and controlled release behavior (properties) are monitored. Antimicrobial activity of these polymeric systems against selected microbial strains are determined.

PROPERTIES OF (METH)ACRYLATE BASED HYDROGELS

The hydrogels based on (meth)acrylate, itaconic acid, and poly(vinyl pyrrolidone) hydrogels are obtained by a simultaneous method, where a single crosslinker has no possibility of any interaction with the second polymer is used. HE(M)A and IA monomeric components are polymerized simultaneously, and in the presence of PVP polymer, in such a way that the HE(M)A/IA copolymer is crosslinked and intermingled with PVP linear polymer. TEMED is used to activate the polymerization process and EGDMA is used as a crosslinker to create a three-dimensional polymeric network. Subsequently, metal ions are embedded in such polymeric network (Scheme 1). These are the procedures used to design antimicrobial hydrogels.

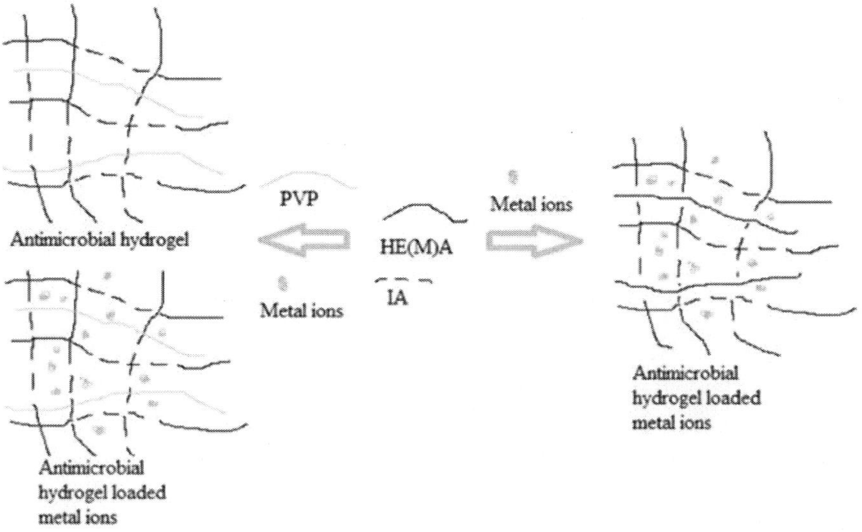

Scheme 1. Schematic of the synthetic route of antimicrobial hydrogels (five series) composed of HE(M)A, IA, and PVP, and metal ions.

Hydrogels based on 2-Hydroxyethyl Methacrylate, Itaconic Acid, and Poly(Vinyl Pyrrolidone) (HIP)

Swelling and Transport Properties of HIP Hydrogels

The pH-sensitive swelling studies of hydrogels based on 2-hydroxyethyl methacrylate, itaconic acid, and poly(vinyl pyrrolidone) (HIP2, HIP5, HIP10 hydrogels) are performed to reveal swelling properties. The samples are swollen in buffers of physiological and pathophysiological values (pH range of 2.20-7.40), at a physiological temperature of 37°C. It can be noticed that all samples exhibit a similar trend of the equilibrium degree of swelling (q_e) on the pH value dependence, typical for anionic hydrogels. The lowest q_e values were detected at a low pH of 2.20. The lowest q_e values are below pK_a of the values of IA groups when the carboxylic groups of itaconic acid are non-ionized. Then, intramolecular hydrogen bonds are formed, resulting in greater network compactness. The ionization occurs when the pH of the medium rises above the pK_a values of both carboxyl groups ($pK_{a1} = 3.85$ and $pK_{a2} = 5.45$). As the ionization intensity increases, the number of constant charge increases, causing increased electrostatic repulsion between the ionizing groups and, consequently, chain separation. This leads to an increase in the hydrophilicity of the network and an increase in the degree of swelling. q_e values increase with increasing of IA content in hydrogels. Thus, the swelling properties of polymeric networks decreased in acidic pH, while a reverse trend was seen in basic pH. However, these hydrogels were found to be hydrolytically stable in phosphate buffer, which makes them be a potential material for biomedical applications (Monleón-Pradas et al. 2001). The swelling kinetic parameters obtained for HIP hydrogel samples showed that Fickian fluid transport takes place within the polymeric network. All PHI samples showed smart pH-sensitive swelling properties.

Equilibrium degree of swelling (q_e) dependence on temperature for PHI, at pH 7.40 was tested. It can be seen that all samples show temperature-sensitive swelling behavior, with a volume phase transition temperature (VPTT) around 41°C, which is in the physiologically important interval. When the temperature is below the VPTT, the polymer exists in a swollen, hydrophilic state. However, as the temperature is raised above the VPTT, the polymer goes through an abrupt conformational rearrangement resulting in a collapsed, hydrophobic state. The conformational change is a result of the break-up of the network of hydrogen-bonded water surrounding the hydrophobic segments of the polymeric chain. This phase transition is more pronounced for the samples with higher PVP content. VPTT value could be tuned to a lower value (body temperature) by adjusting of the IA content in the hydrogel composition.

Antimicrobial Properties of HIP Hydrogels

The assessment of antimicrobial activity is of great importance for biomedical applications. HIP hydrogels samples were tested for three microbial cultures-*E. coli*, *S. aureus*, and *C. albicans*. Based on the results shown in Figure 1, it can be noticed that the antimicrobial effect depends on the fraction of PVP in the hydrogels. As the PVP fraction increases, so does the effectiveness in the fight against microbes. The highest sensitivity for the tested gels was obtained for *C. albicans* pathogenic yeast cells where the percentage of antimicrobial activity was achieved to 87-98%. Slightly lower sensitivity to the antimicrobial activity of PVP hydrogels was shown against the gram-positive *S. aureus* bacterium, with a percentage of the cell number reduction around 60% after the second hour of exposure for the sample with the highest proportion of PVP. The lowest antimicrobial activity had all samples for the Gram-negative bacteria *E. coli*, where a reduction in cell number was in the range of 12-58% after 2 h of exposure. Considering the influence of the exposure time of PVP hydrogels to the indicator strains, an increase in the percentage of cell number reduction with the duration of exposure is observed. This trend is present in all HIP samples and refers to microbial cultures used.

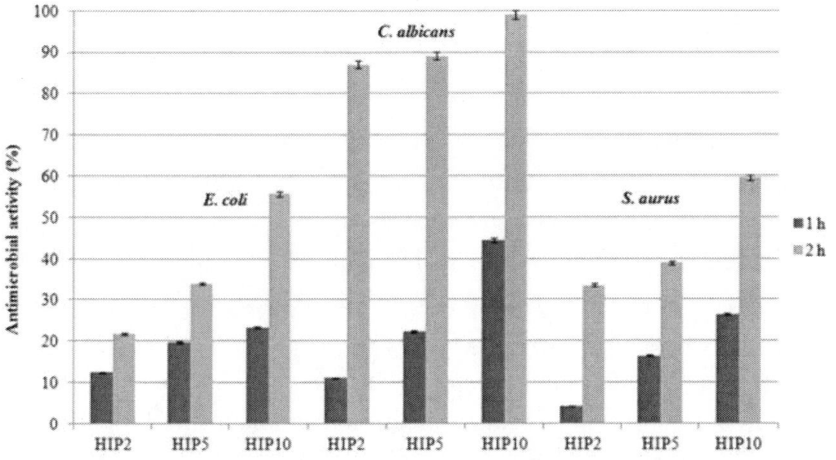

Figure 1. Antimicrobial activity of HIP hydrogels against *E. coli*, *C. albicans*, and *S. aureus*.

Hydrogels Based on 2-Hydroxyethyl Methacrylate, Itaconic Acid and Poly(Vinyl Pyrrolidone) Loaded with Silver(I) Ions (HIPA)

Swelling and Transport Properties of HIPA Hydrogels

The pH-sensitive swelling studies of hydrogels based on 2-hydroxyethyl methacrylate, itaconic acid, and poly(vinyl pyrrolidone) loaded with silver(I) ions (HIP2A, HIP5A, HIP10A hydrogels) are performed to detect swelling properties. The samples are swollen in buffers of physiological and pathophysiological values (pH range of 2.20-7.40), at a physiological temperature of 37 °C. It can be noticed that all samples exhibit a similar trend of the equilibrium degree of swelling (q_e) on the pH value dependence, typical for anionic hydrogels. Results obtained indicated that the incorporation of silver(I) ions leads to a smaller decrease in the degree of swelling, but do not affect the pH-sensitive swelling behavior of the hydrogels based on 2-hydroxyethyl methacrylate, itaconic acid, and poly(vinyl pyrrolidone). The lowest q_e values have samples at a low pH of 2.20. The lowest q_e values are below pK_a of the values of IA groups when the carboxylic groups of itaconic acid are non-ionized. Then, intramolecular hydrogen bonds are formed, resulting in greater network compactness. The ionization occurs when the pH of the medium rises above the pK_a values of both carboxyl groups (pK_{a1} = 3.85 and pK_{a2} = 5.45). As the ionization intensity increases, the number of constant charge increases, causing increased electrostatic repulsion between the ionizing groups and, consequently, chain separation. This leads to an increase in the hydrophilicity of the network and an increase in the degree of swelling. q_e values increase with increasing of IA content in hydrogels. Thus, the swelling properties of polymeric networks decreased in acidic pH, while a reverse trend was seen in basic pH. However, these hydrogels were found to be hydrolytically stable in phosphate buffer, which makes them be a potential material for biomedical applications (Monleón-Pradas et al. 2001). The swelling kinetic parameters obtained for HIPA hydrogel samples showed that Fickian fluid transport takes place within the polymeric network. All PHIA samples showed smart pH-sensitive swelling properties.

Equilibrium degree of swelling (q_e) dependence on temperature for PHIA, at pH 7.40 was tested. All samples showed temperature-sensitive swelling behavior, with a volume phase transition temperature (VPTT) around 41.5 °C, which is in the physiologically important interval. When the temperature is below the VPTT, the polymer exists in a swollen, hydrophilic state. This phase transition is more pronounced for the samples with higher PVP content. The VPTT value can be set to a lower value (body temperature) by adjusting the IA content in the hydrogel composition.

Antimicrobial Properties of HIPA Hydrogels

Antimicrobial activity of HIP2A, HIP5A, HIP10A hydrogels were examined against Gram-negative bacterium *Escherichia coli*, Gram-positive bacterium *Staphylococcus aureus* and yeast *Candida albicans*, which are the most common causes of infection. The study was carried out after one- and two-hour treatment of the cultures of the aforementioned pathogens (Figure 2). The highest sensitivity for the tested gels was obtained for *C. albicans* pathogenic yeast cells where the percentage of antimicrobial activity was achieved up to 93% (Figure 2). Lower sensitivity to the antimicrobial activity of HIPA hydrogels was shown by the gram-positive *S. aureus* bacterium, with a percentage of the cell number reduction of about 72% after the second hour of exposure for the sample with the highest fraction of PVP. The lowest antimicrobial activity had all samples for the Gram-negative bacteria *E. coli*, where a reduction in cell number was lower than 20%. In this series of PHIA hydrogels, the proportion of IA is constant, and therefore the proportion of silver ions that are incorporated. The proportion of PVP was varied. Therefore, silver ions and PVP together give antimicrobial activity, with the proportion of PVP determining the intensity of antimicrobial activity.

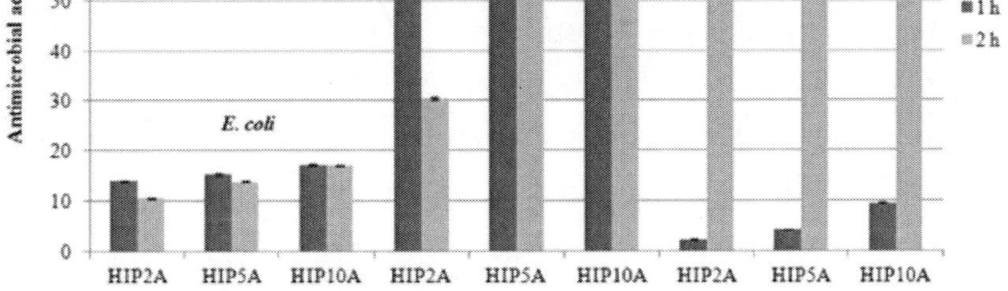

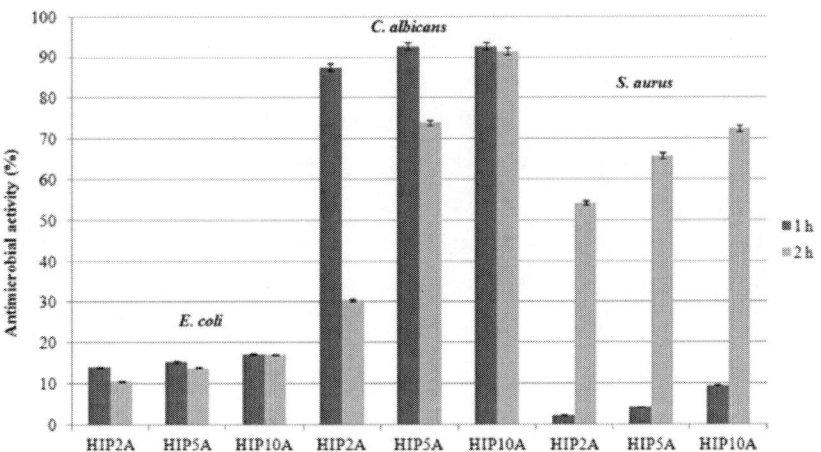

Figure 2. Antimicrobial activity of HIPA hydrogels for *E.coli*, *C. albicans*, and *S. aureus*.

Hydrogels Based on 2-Hydroxyethyl Acrylate and Itaconic Acid Loaded with Silver(I) Ions (A Series)

Swelling and Transport Properties of A Hydrogel Series

Silver(I) ions were incorporated into the hydrogels based on 2-hydroxyethyl acrylate and itaconic acid (ions-free hydrogel series H1-H7 and ions loaded hydrogel series A1-A7). The swelling of hydrogels was monitored over a wide range of physiological fluids (pH of 2.20-7.40), at a temperature of 37 °C, to determine the mode of transport of the fluid through the polymeric network. Due to the presence of itaconic acid, the hydrogels exhibit typical swelling as anionic hydrogels, swelling increases as the pH value increases. Embedded silver ions slightly reduce swelling, but the trend of the dependence of the degree of swelling on the pH value is the same as for ions-free hydrogels. A hydrogel series showed pH-sensitive behavior. The swelling kinetic parameter of A hydrogel series showed that Fickian fluid transport takes place within the polymeric network.

Antimicrobial Properties of A Hydrogel Series

The results obtained from the microbial assay revealed the antibacterial activity of the unloaded and silver loaded hydrogels against methicillin sensitive (MSSA) and methicillin resistant *S. aureus* (MRSA) strains (Figure 3). Loaded hydrogels exhibited higher antibacterial potential against MSSA compared to the unloaded hydrogels. Also, the hydrogel composition influenced bacterial inhibition. By increasing the itaconic acid content, the inhibition of bacterial growth was more pronounced, thus the best results showed hydrogels with 7 mol% of IA against both MSSA and MRSA. In the case of methicillin-resistant strain, silver-free hydrogels were more efficient than silver loaded hydrogels. As it was discussed earlier, the mechanism of silver action against microorganisms can be interpreted in several ways, so this phenomenon is most certainly a consequence of different silver interactions with genetically distinctive cells of MSSA and MRSA.

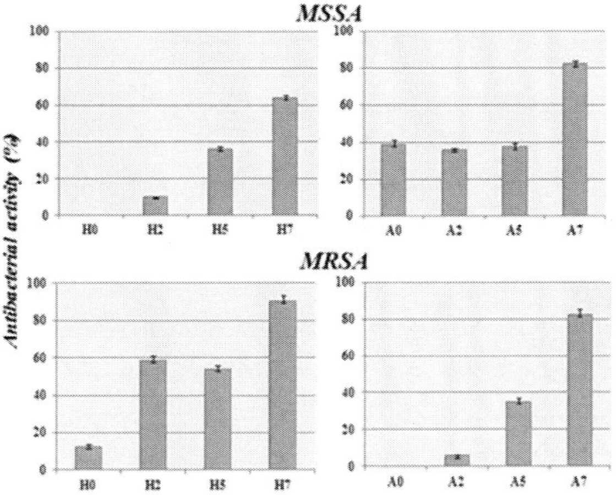

Figure 3. Antibacterial activity of H and A hydrogel series for MSSA and MRSA.

Hydrogels Based on 2-Hydroxyethyl Acrylate And Itaconic Acid Loaded with Copper(II) Ions (C Series)

Swelling and Transport Properties of C Hydrogel Series

Copper(II) ions were incorporated into the hydrogels based on 2-hydroxyethyl acrylate and itaconic acid (ions-free hydrogel series H1-H7 and ions loaded hydrogel series C1-C7). The swelling of hydrogels was monitored over a wide range of physiological fluids (pH of 2.20-7.40), at a temperature of 37 °C, to determine the mode of transport of the fluid through the polymeric network. Due to the presence of itaconic acid, the hydrogels exhibit typical swelling as anionic hydrogels, swelling increases as the pH value increases. Embedded copper ions slightly increase swelling, but the trend of the dependence of the degree of swelling on the pH value is the same as for ions-free hydrogels. C hydrogel series showed pH-sensitive behavior. The swelling kinetic parameter of C hydrogel series showed that Fickian fluid transport takes place within the polymeric network.

Antimicrobial Properties of C Hydrogel Series

The study of controlled release of copper(II) ions from H hydrogel series (hydrogels based on 2-hydroxyethyl acrylate and itaconic acid, samples H0-

H7) in a buffer of pH 7.40 at 37 °C was performed (Figure 4). From the results obtained, it was shown that the release profiles of H hydrogel series depend on the fraction of IA and therefore the incorporated copper ions. The sample with the highest IA content swells the most in the physiological medium and the fastest release of copper ions. Samples with lower IA content swell less, so they release copper(II) ions more slowly (copper ions content is lower because there is less IA to which they are bound).

Antimicrobial activity against three strains Gram-positive bacteria *S. aureus,* Gram-negative bacteria *E. coli*, and fungus *C. albicans* were monitored during the controlled release process. The results obtained (Figure 4) indicate the strong antimicrobial potential of the hydrogels loaded with Cu^{2+} (series C). The inhibition of microbial growth was near 100% against all three strains Gram-positive bacteria *S. aureus,* Gram-negative bacteria *E. coli*, and fungus *C. albicans*. Even the microbial reduction was high, a slight decrease in antimicrobial activity was noticed in regards to selected time points of Cu^{2+} release. Maximum was detected after the first hour of release and minimum after 72 hours, owing to the quantity of residual copper, which actually interacts with microbial cells. From the results, it can be said that by adjusting the content of IA in hydrogels and therefore the embedded copper ions, the antimicrobial activity can be tuned.

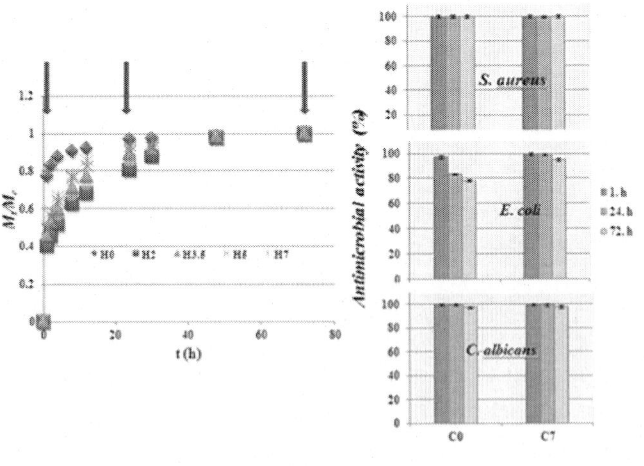

Figure 4. Antimicrobial activity of C hydrogel series during the controlled copper ions release process from H hydrogel series for *E. coli*, *C. albicans*, and *S. aureus*.

Microbial Penetration of C Hydrogel Series

Ideally, hydrogels intended to be used as components of medical devices should not allow the pass of microorganisms through their matrix. However, the likelihood of contamination by Gram-negative bacteria like *Pseudomonas aeruginosa* can be potentially quite high. *P. aeruginosa* is a common environmental Gram-negative bacillus which is one of the most frequent causes of infections in clinical conditions. As it can be noticed from Figure 5, which presents the tested C7 sample hydrogel, there was not any bacterial growth underneath the hydrogel, as well on the upper side. The change in coloration is evidence of the diffusion of the culture medium throughout the hydrogel. Results indicate good barrier characteristics of the hydrogels against one of the most common pathogens nowadays.

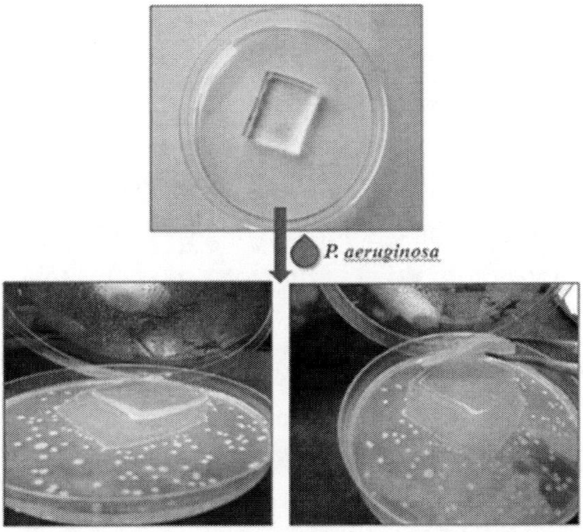

Figure 5. *Pseudomonas aeruginosa* penetration test for C7 sample hydrogel.

Hydrogels Based on 2-Hydroxyethyl Acrylate and Itaconic Acid Loaded with Zinc(II) Ions (Z series)

Swelling and Transport Properties of Z hydrogel Series

Zinc(II) ions were incorporated into the hydrogels based on 2-hydroxyethyl acrylate and itaconic acid (ions-free hydrogel series H1-H7 and ions loaded hydrogel series Z1-Z7). The swelling of hydrogels was monitored over a wide range of physiological fluids (pH of 2.20-7.40), at a temperature of 37 °C, to determine the mode of transport of the fluid through the polymeric network. Due to the presence of itaconic acid, the hydrogels exhibit typical swelling as anionic hydrogels, swelling increases as the pH value increases. Embedded zinc(II) ions slightly increase swelling, but the trend of the dependence of the degree of swelling on the pH value is the same as for ions-free hydrogels. Z hydrogel series showed pH-sensitive behavior. The swelling kinetic parameter of Z hydrogel series showed that Fickian fluid transport takes place within the polymeric network.

Antimicrobial Properties of Z hydrogel Series

The study of controlled release of zinc(II) ions from H hydrogel series (hydrogels based on 2-hydroxyethyl acrylate and itaconic acid, samples H0-H7) in a buffer of pH 7.40 at 37 °C was performed (Figure 6). From the results obtained, it was shown that the release profiles of H hydrogel series depend on the fraction of IA and therefore the loaded zinc ions. The sample with the highest IA content swells the most in the physiological medium and the fastest release of zinc ions. Samples with lower IA content swell less, so they release zinc ions more slowly (zinc ions content is lower because there is less IA to which they are bound).

Antimicrobial activity against E. coli microbe was monitored during the controlled release process. The results obtained showed that the inhibition of *E. coli* growth was reached over 90% when the microbial cells were exposed to all tested Z hydrogel series (Figure 6). In regards to selected Zn^{2+} release time points, the antibacterial potential decreased. Therefore it can be said that the maximum of microbial inhibition capacity hydrogels exhibited at the beginning of Zn^{2+} release monitoring, and the lowest at the end of the study. As it was discussed previously, this behavior is related to the activity of residual zinc in hydrogels after the release started. From the results, it can be said that by adjusting the content of IA in hydrogels and therefore the embedded zinc ions, the antimicrobial activity can be tuned.

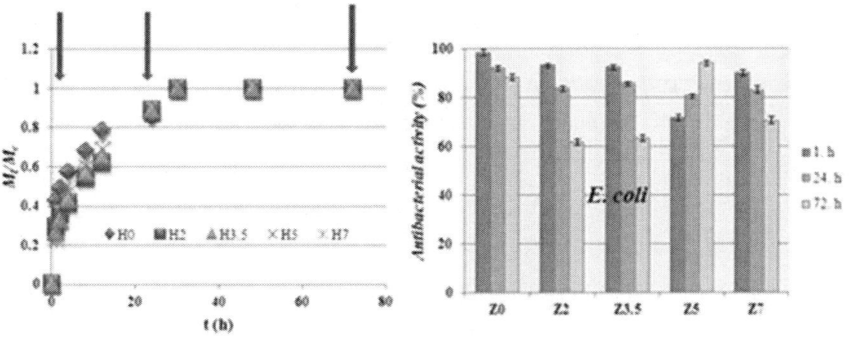

Figure 6. Antimicrobial activity of Z hydrogel series during the controlled zinc ions release process from H hydrogel series for *E. coli.*

CONCLUSION

Polymeric networks of hydrogels are very specific 3-D structures and allow fluid circulation through the network. These transport properties provide the network to adapt to the circumstances imposed in the context of changes in pH and temperature. Such structures can also engage in the fight against microbes.

Hydrogels as antibacterial biomaterials can be an alternative and amenable solution to traditional antibiotic treatments. Favorable swelling, controlled and prolonged release, local administration, stimulated switch on-off release, enhanced mechanical strength, and improved biocompatibility are all important advantages that a broad diversity of hydrogels can provide and that is exactly what antibacterial biomaterials currently require. Based on these factors, intelligent hydrogel platforms should be exploited to overcome the challenges of local antibacterial drugs. Antibacterial biomaterials, their unique combinations, and the approaches currently being developed will provide a promising future for anti-infection treatment.

In our review, five series of polymeric hydrogel systems, based on monomers of 2-hydroxyethyl methacrylate, 2-hydroxyethyl acrylate, and itaconic acid, with poly(N-vinylpyrrolidone), and medicative metal ions. All samples showed intelligent behavior and properties that are extremely favorable for biomedical applications-swelling, controlled metal ions release, and powerful antimicrobial potential.

The antimicrobial activity of these HIP samples depends on the PVP fraction in the hydrogels and how the PVP fraction increases, thus increasing the efficiency. The best antimicrobial properties were shown against *C. albicans*, the inhibition of bacterial growth was almost complete. The lower hydrogel activity was detected against *S. aureus* strain. The largest inhibition for the sample with a maximum of PVP content was 60%. The lowest antimicrobial activity had all samples against Gram-negative bacteria *E. coli*. In the investigated period of antimicrobial activity, inhibition of bacterial growth increased with time, for all samples and bacterial strains.

Both antimicrobial agents, PVP, and silver(I) ions, contribute to the antimicrobial activity of HIPA hydrogels. The highest sensitivity of HIPA hydrogels was obtained for *C. albicans*, where the percentage of antimicrobial activity was achieved up to 93%. Lower sensitivity to the antimicrobial activity of HIPA hydrogels was shown against Gram-positive *S. aureus*, with a percentage of the cell number reduction of about 72% after the second hour of exposure for the sample with the highest fraction of PVP. The lowest antimicrobial activity had all samples against Gram-negative bacteria *E. coli*, where a reduction in cell number was lower than 20%.

The microbial assay for A hydrogel series revealed the antibacterial activity of the unloaded and silver loaded hydrogels against both *S. aureus* strains. Loaded hydrogels exhibited higher antibacterial potential against MSSA compared to the unloaded hydrogels. Also, the hydrogel composition influenced bacterial inhibition. By increasing the itaconic acid content, the inhibition of bacterial growth was more pronounced, thus the best results showed hydrogels with 7 mol% of IA against both MSSA and MRSA. In the case of methicillin-resistant strain, silver-free hydrogels were more efficient than silver loaded hydrogels.

Antimicrobial activity of C hydrogel series against three strains Gram-positive bacteria *S. aureus,* Gram-negative bacteria *E. coli*, and fungus *C. albicans* were monitored during the controlled release process. The obtained results indicate the strong antimicrobial potential of the hydrogels loaded with Cu^{2+} against all three microbial strains. Results of a microbial penetration test against *Pseudomonas aeruginosa* indicate good barrier characteristics of the hydrogels.

Antimicrobial activity of Z hydrogel series against Gram-negative bacteria *E. coli* was monitored during the controlled Zn^{2+} ions release process. The results obtained showed that the inhibition of *E. coli* growth was reached over 90%. It can be said that the maximum of microbial inhibition capacity hydrogels exhibited at the beginning of Zn^{2+} release monitoring, and the lowest at the end of the study. From these results, it can be said that by adjusting the content of IA in hydrogels and therefore the embedded metal ions, the antimicrobial activity can be tuned.

ACKNOWLEDGMENTS

This work was supported by the Ministry of Education, Science and Technological Development of the Republic of Serbia (Contracts No 451-03-68/2020-14/172062 and 451-03-68/2020-14/172026).

REFERENCES

Abd El-Mohdy, H. L. 2013. Radiation synthesis of nanosilver/poly vinyl alcohol/cellulose acetate/gelatin hydrogels for wound dressing. *Journal of Polymer Research* 20(177):1-12.

Abdelrazek, E. M., Ragab, H. M., and Abdelaziz, M. 2013. Physical characterization of poly(vinyl pyrrolidone) and gelatin blend films doped with magnesium chloride. *Plastic and Polymer Technology (PAPT)* 2:1-8.

Agarwal, Happy, Menon, Soumya, Venkat Kumar, S., and Rajeshkumar S. 2018. Mechanistic study on antibacterial action of zinc oxide nanoparticles synthesized using green route. *Chemico-Biological Interactions* 286:60–70.

Ahmed, Enas M. 2015. Hydrogel: Preparation, characterization, and applications: A review. *Journal of Advanced Research* 6:105-121.

Ajji, Z., Othman, I., and Rosiak, J. M., 2005. Production of hydrogel wound dressings using gamma radiation. *Nuclear Instruments and Methods in Physics Research Section B: Beam Interactions with Materials and Atoms* 229:375–380.

Ashfaq, Mohammad, Verma, Nishith, and Khan, Suphiya 2016. Copper/zinc bimetal nanoparticles-dispersed carbon nanofibers: A novel potential antibiotic material. *Materials Science and Engineering: C* 59:938–947.

Babić, Marija M., Antić, Katarina M., Jovašević Vuković, Jovana S., Božić, Bojan Dj., Davidović, Slađana, Filipović, Jovanka M., and Tomić, Simonida Lj. 2015. Oxaprozin/poly(2-hydroxyethyl acrylate/itaconic acid) hydrogels: morphological, thermal, swelling, drug release and antibacterial properties. *Journal of Material Science* 50:906-922.

Baino, Francesco 2010. The use of polymers in the treatment of retinal detachment: Current trends and future perspectives. *Polymers* 2:286-322.

Bajpai, A. K., Shukla, Sandeep K., Bhanu, Smitha, and Kankane, Sanjana 2008. Responsive polymers in controlled drug delivery. *Progress in Polymer Science* 33:1088–1118.

Barrett, Graham D., Constable, Ian J., and Stewart, Andrew D. 1986. Clinical results of hydrogel lens implantation. *Journal of Cataract and Refractive Surgery* 12:623-31.

Barros, Janaina A. G., Brant, Antonio J. C., and Catalani, Luiz H. 2011. Hydrogels from chitosan and a novel copolymer poly(N-vinyl-2-pyrrolidone-co-acrolein). *Materials Sciences and Applications* 2:1058-1069.

Bera, Rabin, Dey, Ayan, and Chakrabarty, Debabrata 2015. Synthesis, characterization, and drug release study of acrylamide-co-itaconic acid based smart hydrogel. *Polymer Engineering and Science* 55:113-122.

Bhowmick, Sirsendu, and Koul, Veena 2016. Assessment of PVA/silver nanocomposite hydrogel patch as antimicrobial dressing scaffold: synthesis, characterization and biological evaluation. *Materials Science and Engineering: C* 59:109–119.

Biazar, Esmaeil, Roveimiab, Ziba, Shahhosseini, Gholamreza, Khataminezhad, Mohammadreza, Zafari, Mandana, and Majdi, Ali

2012. Biocompatibility evaluation of a new hydrogel dressing based on polyvinylpyrrolidone/polyethylene glycol. *Journal of Biomedicine and Biotechnology* 2012:1-5.

Boehle, Katherine E., Gilliand, Jake, Wheeldon, Christopher R., Holder, Amethyst, Adkins, Jaclyn A., Geiss, Brian J., Ryan, Elizabeth P., and Henry, Charles S. 2017. Utilizing paper-based devices for antimicrobial-resistant bacteria detection. *Angewandte Chemie* 56:6886-6890.

Caliari, Steven R., and Burdick, Jason A. 2016. A practical guide to hydrogels for cell culture. *Nature Methods* 13:405-414.

Chadha, Renu, Kapoor, V. K., and Kumar, Amit 2006. Analytical techniques used to characterize drug- polyvinylpyrrolidone systems in solid and liquid states-An overview. *Journal of Scientific and Industrial Research* 65:459-469.

Chai, Qinyuan, Jiao, Yang, and Yu, Xinjun 2017. Hydrogels for biomedical applications: their characteristics and the mechanisms behind them. *Gels* 3:1-15.

Chen, Su, Hu, Ting, Tian, Yuan, Chen, Li, and Pojman, John A. 2007. Facile synthesis of poly(hydroxyethyl acrylate) by frontal free-radical polymerization. *Journal of Polymer Science and Polymer Chemistry* 45:873-881.

Chirila, Traian and Harkin, Damien 2016. *Biomaterials and Regenerative Medicine in Ophthalmology*. 2nd ed. UK: Elsevier.

Clayton, Anthony B., Chirila, Traian V., and Lou, Xia 1997. Hydrophilic sponges based on 2-hydroxyethyl methacrylate. Effect of crosslinking agent reactivity on mechanical properties. *Polymer International* 44:201-207.

Cometa, Stefania, Iatta, Roberta, Ricci, Maria A., Ferretti, Concetta, and De Giglio, Elvira 2013. Analytical characterization and antimicrobial properties of novel copper nanoparticle–loaded electrosynthesized hydrogel coatings. *Journal of Bioactive and Compatible Polymers* 28:508–522.

Das, Nilimanka 2013. Preparation methods and properties of hydrogel: A review. *International Journal of Pharmacy and Pharmaceutical Sciences* 5: 112-117.

De Silva, Awanhi D., Hettiarachchi, Buddhika U., Nayanajith, L. D. C., Yoga Milani, M. D., and Motha, J. T. S. 2011. Development of a PVP/kappa-carrageenan/PEG hydrogel dressing for wound healing applications in Sri Lanka. *Journal of the National Science Foundation of Sri Lanka* 39:25-33.

Dharmalingam, Koodalingam, Bordoloi, Devivasha, Kunnumakkara, Ajaikumar B., and Anandalakshmi, Ramalingam 2020. Formation and characterization of zinc oxide complexes in composite hydrogel films for potential wound healing applications. *Polymer Composites* doi: 10. 1002/pc.25538.

Domingues, Joana A., Bonelli, Nicole, Giorgi, Rodorico, Fratini, Emiliano, Gorel, Florence, and Baglioni, Piero 2013. Innovative hydrogels based on semi-interpenetrating p(HEMA)/PVP networks for the cleaning of water-sensitive cultural heritage artifacts. *Langmuir* 29:2746-2755.

Erizal, Tjahyono, Dian, P. P., and Darmawan 2013. Synthesis of polyvinyl pyrrolidone (PVP)/κ-carrageenan hydrogel prepared by gamma radiation processing as a function of dose and PVP concentration. *Indonesian Journal of Chemistry* 13:41-46.

Fan, Zengjie, Liu, Bin, Wang, Jinqing, Zhang, Songying, Lin, Qianqian, Gong, Peiwei, Ma, Limin, and Yang, Shengrong 2014. A novel wound dressing based on Ag/graphene polymer hydrogel: Effectively kill bacteria and accelerate wound healing. *Advanced Functional Materials* 24:3933-3943.

García-Barrasa, Jorge, López-de-Luzuriaga, Jose M., and Monge, Miguel 2010. Silver nanoparticles: synthesis through chemical methods in solution and biomedical applications. *Central European Journal of Chemistry* 9:7–19.

Gehrke, S. H., Biren, D., Hopkins, J. J. 1995. Evidence for Fickian water transport in initially glassy poly(2-hydroxyethyl methacrylate). *Journal of Biomaterials Science Polymer Edition* 6:375-390.

Gils, Palapparambil S., Sahu, Nalin K., Ray, Debajyoti, and Sahoo, Prafulla K. 2011. Hydrolyzed collagen-based hydrogel system: design, characterization, and application in drug delivery. *International Journal of Macromolecular Science* 1:1-8.

Giri, Nabaraj, Natarajan, R. K., Gunasekaran, Sethu, and Shreemathi, S. 2011. ^{13}C NMR and FTIR spectroscopic study of blend behavior of PVP and nanosilver particles. *Archives of Applied Science Research* 3: 624-630.

Global Burden of Disease Study 2013 Collaborators 2015. Global, regional, and national incidence, prevalence, and years lived with disability for 301 acute and chronic diseases and injuries in 188 countries, 1990–2013: a systematic analysis for the Global Burden of Disease Study 2013. *Lancet* 386(9995):743–800.

González-Sánchez, Isabel M., Perni, Stefano, Tommasi, Giacomo, Morris, Nathanael G., Hawkins, Karl, López-Cabarcos, Enrique, and Prokopovich, Polina 2015. Silver nanoparticle based antibacterial methacrylate hydrogels potential for bone graft applications. *Materials Science and Engineering: C*. 5:332-340.

Gopal, Anu, Kant, Vinay, Gopalakrishnan, Anu, Tandan, Surendra K., and Kumar, Dinesh 2014. Chitosan-based copper nanocomposite accelerates healing in excision wound model in rats. *European Journal of Pharmacology* 731:8–19.

Gould, Simon W. J., Fielder, Mark D., Kelly, Alison F., Morgan, Marina, Kenny, Jackie, Naughton, and Declan P. 2009. The antimicrobial properties of copper surfaces against a range of important nosocomial pathogens. *Annals of Microbiology* 59:151-156.

Grass, G., Rensing, C., Solioz, M. 2011. Metallic copper as an antimicrobial surface. *Applied and Environmental Microbiology* 77:1541–1547.

Gulrez, S. K. H., Al-Assaf, S., and Phillips, G. O. 2011. Progress in molecular and environmental bioengineering - from analysis and modeling to technology applications. In *Hydrogels: Methods of Preparation, Characterisation and Applications*, edited by Angelo Carpi. IntechOpen, 117-150.

Guo, Chuigen, Zhou, Lin, and Jianxiong, Lv 2013. Effects of expandable graphite and modified ammonium polyphosphate on the flame-retardant and mechanical properties of wood flourpolypropylene composites. *Polymer Composites* 21:449-456.

Guo, Liya, Yuan, Weiyong, Lu, Zhisong, Li, Chang M. 2013. Polymer/nanosilver composite coatings for antibacterial applications. *Colloids and Surfaces A: Physicochemical and Engineering Aspects* 439:69–83.

Halpenny, Genevieve M., Steinhardt, Rachel C., Okialda, Krystle A., Mascharak, and Pradip K. 2009. Characterization of pHEMA-based hydrogels that exhibit light-induced bactericidal effect via release of NO. *Journal of Materials Science: Materials in Medicine* 20:2353-2360.

Harris, Edward D., Rayton, John K., Balthrop, James E., Di Silvestro, Robert A., and Garcia-de-Quevedo, Margaret 1980. Copper and the synthesis of elastin and collagen. *Ciba Found Symposium* 79:163-182.

He, Jingwei, Söderling, Eva, Lassila, Lippo V. J., and Vallittu, Pekka K. 2012. Incorporation of an antibacterial and radiopaque monomer in to dental resin system. *Dental Materials Journal* 28:110-117.

Himi, Madolia 2013. Preparation and evaluation of stomach specific IPN hydrogels for oral drug delivery: A review. *JDDT* 3:131-140.

Hoffman, Allan S. 2012. Hydrogels for biomedical applications. *Advanced Drug Delivery Reviews* 64:18–23.

Hu, Guo-fu 1998. Copper stimulates proliferation of human endothelial cells under culture. *Journal of Cellular Biochemistry* 69:326–335.

Huh, Ae J., Kwon, and Young J. 2011. Nanoantibiotics: A new paradigm for treating infectious diseases using nanomaterials in the antibiotics resistant era. *Journal of Controlled Release* 156:128–145.

Ingle, Avinash P., Duran, Nelson, and Rai, Mahendra 2014. Bioactivity, mechanism of action, and cytotoxicity of copper-based nanoparticles: a review. *Applied Microbiology and Biotechnology* 98:1001–1009.

Jaiswal, Maneesh, Koul, Veena, and Dinda, Amit K. 2016. In vitro and in vivo investigational studies of a nanocomposite-hydrogel-based dressing with a silver-coated chitosan wafer for full-thickness skin wounds. *Journal of Applied Polymer Science* 133:(43472)1-12.

Ji, Haiwei, Sun, Hanjun, and Qu, Xiaogang 2016. Antibacterial applications of graphene-based nanomaterials: Recent achievements and challenges. *Advanced Drug Delivery Reviews*. 105:176-189.

Jiao, Yanpeng, Liu, Zonghua, Ding, Shan, Li, Lihua, and Zhou, Changren 2006. Preparation of biodegradable crosslinking agents and application in PVP hydrogel. *Journal of Applied Polymer Science* 101:1515–1521.

Jones, Nicole, Ray, Binata, Ranjit, Koodali T., and Manna, Adhar C. 2008. Antibacterial activity of ZnO nanoparticle suspensions on a broad spectrum of microorganisms. *FEMS Microbiology Letters* 279:71–76.

Klinkajon, Wimonwan, and Supaphol, Pitt 2014. Novel copper (II) alginate hydrogels and their potential for use as anti-bacterial wound dressings. *Biomedical Materials* 9:1-11.

Kong, Hyeyoung, and Jang, Jyongsik 2008. Antibacterial properties of novel poly(methyl methacrylate) nanofiber containing silver nanoparticles. *Langmuir* 24:2051-2056.

Kruk, Tomasz, Szczepanowicz, Krzysztof, Stefanska, Joanna, Socha, Robert P., and Warszynski, Piotr 2015. Synthesis and antimicrobial activity of monodisperse copper nanoparticles. *Colloids and Surfaces B: Biointerfaces* 128:17–22.

La, Young H., McCloskey, Bryan D., Sooriyakumaran, Ratnam, Vora, Ankit, Freeman, Benny, Nassar, Majed, Hedrick, James, Nelson, Alshakim, and Allen, Robert 2011. Bifunctional hydrogel coatings for water purification membranes: Improved fouling resistance and antimicrobial activity. *Journal of Membrane Science* 372:285-291.

Lansdown, Alan B. G., Mirastschijski, Ursula, Stubbs, Nicky, Scanlon, Elizabeth, and Agren, Magnus S. 2007. Zinc in wound healing: theoretical, experimental, and clinical aspects. *Wound Repair and Regeneration* 15:2–16.

Lara, Humberto H., Ayala-Núñez, Nilda V., del Carmen Ixtepan Turrent, Liliana, and Padilla, Cristina R. 2010. Bactericidal effect of silver nanoparticles against multidrug-resistant bacteria. *World Journal of Microbiology and Biotechnology* 26:615–621.

Le Ouay, Benjamin, and Stellacci, Francesco 2015. Antibacterial activity of silver nanoparticles: A surface science insight. *Nano Today* 10:339–354.

Li, Yong, and Tanaka, Toyoichi 1990. Kinetics of swelling and shrinking of gels. *Journal of Chemical Physics* 92:1365-1371.

Lipsky, Benjamin A., and Hoey, Christopher 2009. Topical antimicrobial therapy for treating chronic wounds. *Clinical Infectious Disease* 49:1541–1549.

Liu, Lin, Feng, Xiangru, Pei, Yueting, Wang, Jinze, Ding, Jianxun, and Chen, Li 2018. α-Cyclodextrin concentration-controlled thermo-sensitive supramolecular hydrogels. *Materials Science and Engineering: C* 82:25-28.

Liu, Y., He, L., Mustapha, A., Li, H., Hu, Z. Q., and Lin, M. 2009. Antibacterial activities of zinc oxide nanoparticles against Escherichia coli O157:H7. *Journal of Applied Microbiology* 107:1193-201.

Malmsten, Martin 2011. Antimicrobial and antiviral hydrogels. *Soft Matter* 7:8725–8736.

Malmsten, Martin 2014. Nanomaterials as Antimicrobial Agents in *Handbook of Nanomaterials Properties*, edited by Bharat Bhushan, Dan Luo, Stefan Zauscher, Wolfgang Sigmund, and Scott R. Schricker, 1053-1075. Berlin: Springer.

Marsano, E., Bianchi, E., Vicini, S., Compagnino, L., Sionkowska, A., Skopińska, J., and Wiśniewski, M. 2005. Stimuli responsive gels based on interpenetrating network of chitosan and poly(vinylpyrrolidone). *Polymer* 46:1595-1600.

Menapace, Rupert, Skorpik, Christian, Juchem, Muriel, Scheidel, Wolfgang, and Schranz, R. 1989. Evaluation of the first 60 cases of poly HEMA posterior chamber lenses implanted in the sulcus. *Journal of Cataract and Refractive Surgery* 15:264–271.

Mikolay, Andre, Huggett, Susanne, Tikana, Ladji, Grass, Gregor, Braun, Jorg, and Nies, Dietrich H. 2010. Survival of bacteria on metallic copper surfaces in a hospital trial. *Applied Microbiology and Biotechnology* 87:1875–1879.

Molton, James S., Tambyah, Paul A., Ang, Brenda S., Ling, Moi L., and Fisher, Dale A. 2013. The global spread of healthcare-associated multidrug-resistant bacteria: a perspective from Asia. *Clinical Infectious Diseases* 56:1310-1318.

Monleón-Pradas, M., Gómez-Ribelles, J. L., Serrano-Aroca, Á., Gallego-Ferrer, G., Suay-Antón, J., and Pissis, P. 2001. Interaction between water and polymer chains in poly(hydroxyethyl acrylate) hydrogels. *Colloid and Polymer Science* 279:323-330.

Monleón-Pradas, M., Gómez-Ribelles, J. L., Serrano-Aroca, Á., Gallego-Ferrer, G., Suay-Antón, J., and Pissis, P. 2001. Porous poly(2-hydroxyethyl acrylate) hydrogels. *Polymer* 42:4667-4674.

Moszner, Norbert, and Salz, Ulrich 2001. New developments of polymeric dental composites. *Progress in Polymer Science* 26:535-576.

Naghdeali, Mokhtar H., and Adimi, Maryam 2015. Comparison between acrylic acid and methacrylamide on release and swelling properties for hydrogels based on PVP. *Biological Forum – An International Journal* 7:304-308.

Ng, Victor W. L., Chan, Julian M. W., Sardon, Haritz, Ono, Robert J., García, Jeannette M., Yang, Yi Y., and Hedrick, James L. 2014. Antimicrobial hydrogels: a new weapon in the arsenal against multidrug-resistant infections. *Advanced Drug Delivery Reviews* 78:46-62.

Okabe, Mitsuyasu, Lies, Dwiarti, Kanamasa, Shin, and Park, Enoch Y., 2009. Biotechnological production of itaconic acid and its biosynthesis in *Aspergillus terreus. Applied Microbiology and Biotechnology* 84:597-606.

Okay, Oguz 2009. General Properties of Hydrogels. In *Hydrogel Sensors and Actuators, Springer Series on Chemical Sensors and Biosensors*, edited by Gerald Gerlach and Karl-Friedrich Arnd, 1-14. Berlin Heidelberg: Springer-Verlag.

Pal, K., Banthia, A. K., and Majumdar, D. K. 2009. Polymeric Hydrogels: Characterization and Biomedical Applications –A mini-review. *Designed Monomers and Polymers* 12:197-220.

Parsons, Carole, McCoy, Colin P., Gorman, Sean P., Jones, David S., Bell, Steven E., Brady, Clare, and McGlinchey, Seana M. 2009. Anti-infective photodynamic biomaterials for the prevention of intraocular lens-associated infectious endophthalmitis. *Biomaterials* 30:597-602.

Pelgrift, Robert Y., and Friedman, Adam J. 2013. Nanotechnology as a therapeutic tool to combat microbial resistance. *Advanced Drug Delivery Reviews* 65(13–14):1803–1815.

Peppas, N. A., Bures, P., Leobandug, W., Ichikawa, H. 2000. Hydrogels in pharmaceutical formulations. *European Journal of Pharmaceutics and Biopharmaceutics* 50:27-46.

Percival, Steven L., Slone, Will, Linton, Sara, Okel, Tyler, Corum, Linda, and Thomas, John G. 2011. The antimicrobial efficacy of a silver alginate dressing against a broad spectrum of clinically relevant wound isolates. *International Wound Journal* 8:237–243.

Petruccioli, M., Pulci, V., Federici, F. 1999. Itaconic acid production by *Aspergillus terreus* on raw starchy materials. *Letters in Applied Microbiology* 28:309–312.

Ping, Xiang, Wang, Mozhen, and Xuewu, Ge 2011. Surface modification of poly(ethylene terephthalate) (PET) film by gamma-ray induced grafting of poly(acrylic acid) and its application in antibacterial hybrid film. *Radiation Physics and Chemistry* 80:567-572.

Prasitsilp, M., Siriwittayakorn, T., Molloy, R., Suebsanit, N., Siriwittayakorn, P., Veeranondha, S. 2003. Cytotoxicity study of homopolymers and copolymers of 2-hydroxyethyl methacrylate and some alkyl acrylates for potential use as temporary skin substitutes. *Journal of Materials Science: Materials in Medicine* 14:595-600.

Ristić, Branko, Popović, Zoran, Adamović, Dragan, and Devedžić, Goran 2011. Izbor biomaterijala u ortopedskoj hirurgiji. *Vojnosanitetski Pregled* 67:847-855.

Rodríguez-Hernández, Juan 2016. Antimicrobial Hydrogels. In *Polymers against microorganisms* 1st edn., 179-204. Springer.

Salomé Veiga, Ana, and Schneider, Joel P. 2013. Antimicrobial hydrogels for the treatment of infection. *Biopolymers* 100:637–644.

Sanna, Roberta, Alzari, Valeria, Nuvoli, Daniele N., Scognamillo, Sergio, Marceddu, Salvatore, and Mariani, Alberto 2012. Polymer hydrogels of 2-hydroxyethyl acrylate and acrylic acid obtained by frontal polymerization. *Journal of Polymer Science Part A: Polymer Chemistry* 50:1515-1520.

Sariri, Reyhaneh, and Jafarian, Vahab 2002. The effect of itaconic acid on biocompatibility of HEMA. *European Cells and Materials* 4:41.

Sarker Satya R., Hossain, M., Polash, Shakil A., Takikawa, Masato, Shubhra, Razib D., Saha, Tanushree, Islam, Zinia, Hossain, S., Hasan, A., Takeoka Shinji, and Sarker, Satya R. 2019. Investigation of the antibacterial activity and in vivo cytotoxicity of biogenic silver nanoparticles as potent therapeutics. *Frontiers in Bioengineering and Biotechnology* 7:1-14.

Serrano-Aroca, Á., Campillo-Fernández, A. J., Gómez-Ribelles, J. L., Monleón-Pradas, M., Gallego-Ferrer, G., Pissis, P. 2004. Porous poly(2-hydroxyethyl acrylate) hydrogels prepared by radical polymerization with methanol as diluent. *Polymer* 45:8949-8955.

Shalaby, Shalaby W., Nagatomi, Sheila D., and Peniston, Shawn J. 2007. Polymeric biomaterials for articulating joint repair and total joint replacement. In *Polymers for dental and orthopedic applications*, edited by Shalaby W. Shalaby, Ulrich Salz, 185-212. Boca Raton, CRC Press.

Shi, Lin, Chen, Jiongrun, Teng, Lijing, Wang, Lin, Zhu, Guanglin, Liu, Sa, Luo, Zhengtang, Shi, Xuetao, Wang, Yingjun, and Ren, Li 2016. The antibacterial applications of graphene and its derivatives. *Small* 12:4165-4184.

Simeon, Alain, Emonard, Herve, Hornebeck, William, and Maquart, F. X. 2000. The tripeptide-copper complex glycyl-L-histidyl-Llysine-Cu^{2+} stimulates matrix metalloproteinase-2 expression by fibroblast cultures. *Life Sciences* 67:2257–2265.

Singh, Garima, Lohani, Alka, and Bhattacharya, Shiv S. 2014. Hydrogel as a novel drug delivery system: a review. *Journal of Fundamental Pharmaceutical Research* 2:35-48.

Singh, Rita, and Singh, Durgeshwer 2012. Radiation synthesis of PVP/alginate hydrogel containing nanosilver as wound dressing. *Journal of Materials Science: Materials in Medicine* 23:2649–2658.

Song, Jooyoung, and Jang, Jyongsik 2014. Antimicrobial polymer nanostructures: Synthetic route, mechanism of action and perspective. *Advances in Colloid and Interface Science* 203:37-50.

Stoy, Vladimir A. 1996. *Hydrogels: Speciality plastics for biomedical and pharmaceutical applications*. Basel: Technomic Publishers.

Straccia, Maria C., d' Ayala, Giovanna G., Romano, Ida, and Laurienzo, Paola 2015. Novel zinc alginate hydrogels prepared by internal setting method with intrinsic antibacterial activity. *Carbohydrate Polymers* 125:103–112.

Sudarkodi, C., Subha, K., Kanimozhi, K., and Panneerselvam, A. 2012. Optimization and production of itaconic acid using *Aspergillus flavus. Advances in Applied Science Research* 3:1126-1131.

Sui, Xiaofeng, Feng, Xueling, Di Luca, Andrea, Van Blitterswijk, Clemens A., Moroni, Lorenzo, Hempenius, Mark A., and Vancso, Julius G. 2013. Poly(N-isopropylacrylamide)–poly(ferrocenylsilane) dual-responsive hydrogels: synthesis, characterization and antimicrobial applications. *Polymer Chemistry* 4:337-342.

Tan, Siang Y., and Tatsumura, Yvonne 2015. Alexander Fleming (1881-1955): Discoverer of penicillin. *Singapore Medical Journal* 56:366-367.

Tanaka, Toyoichi 1981. Gels. *Scientific American* 244:124-138.

Tomar, Ritu, Sharma, and Chirag R. 2013. Fabrication and characterization of conducting polymer composite. *International Journal of Organic Electronics* 2:1-8.

Tomić, Simonida Lj., Mićić, Maja M., Dobić, Sava N., Filipović, Jovanka M., and Suljovrujić, Edin H. 2010. Smart poly (2-hydroxyethyl methacrylate/itaconic acid) hydrogels for biomedical application. *Radiation Physics and Chemistry* 79:643-649.

Tomić, Simonida Lj., Suljovrujić, Edin H., and Filipović, Jovanka M. 2006. Biocompatible and bioadhesive hydrogels based on 2-hydroxyethyl methacrylate, monofunctional poly(alkylene glycol)s and itaconic acid. *Polymer Bulletin* 57:691–702.

Van Blitterswijk, Clemens, De Boer Jan. 2014. *Tissue Engineering*. United States of America: Academic Press.

Van Hoek, Angela H., Mevius Dik, Guerra, Beatriz, Mullany, Peter, Roberts, Adam P., and Aarts, Henk J. 2011. Acquired antibiotic resistance genes: an overview. *Frontiers in Microbioogy* 2:1-27.

Wang, Jianquan, and Wu, Wenhui 2005. Swelling behaviors, tensile properties and thermodynamic studies of water sorption of 2-

hydroxyethyl methacrylate/epoxy methacrylate copolymeric hydrogels. *European Polymer Journal* 41:1143-1151.

Wang, Wenbo, Wang, Qin, and Wang, Aiqin 2011. pH-responsive carboxymethylcellulose-g-poly(sodium acrylate)/polyvinylpyrrolidone semi-IPN hydrogels with enhanced responsive and swelling properties. *Macromolecular Research* 19:57-65.

Wei, Lingyu, Chen, Jinjin, Zhao, Shuhua, Ding, Jianxun, and Chen, Xuesi 2017. Thermo-sensitive polypeptide hydrogel for locally sequential delivery of two-pronged antitumor drugs. *Acta Biomaterialia* 58:44-53.

Wei, Qing B., Fu, Feng, Zhang, Yu Q., Wang, Qiao, and Ren, Yi X. 2014 pH-responsive CMC/PAM/PVP semi-IPN hydrogels for theophylline drug release. *Journal of Polymer Research* 21:1-10.

Whitehead, Kathryn A., Colligon, John, and Verran, Joanna 2005. Retention of microbial cells in substratum surface features of micrometer and sub-micrometer dimensions. *Colloids and Surfaces B: Biointerfaces* 41:129-138.

Xie, Huiqi, and Kang, Y. J. 2009. Role of copper in angiogenesis and its medicinal implications. *Current Medicinal Chemistry* 16:1304-1314.

Xu, Bing, Li, Yongmao, Gao, Fei, Zhai, Xinyun, Sun, Mengge, Lu, William, Cao, Zhiqiang, and Liu, Wenguang 2015. High strength multifunctional multiwalled hydrogel tubes: ion triggered shape memory, antibacterial, and anti-inflammatory efficacies. *ACS Applied Materials & Interfaces* 7:16865−16872.

Yanpeng, Jiao, Zonghua, Liu, Shan, Ding, Lihua, Li, and Changren, Zhou 2006 Preparation of biodegradable crosslinking agents and application in PVP hydrogel. *Journal of Applied Polymer Science* 101:1515-1521.

Zheng, Yuhao, Cheng, Yilong, Chen, Jinjin, Ding, Jianxun, Li, Mingqiang, Li, Chen, Wang, Jin C., and Chen Xuesi 2017. Injectable hydrogel-microsphere construct with sequential degradation for locally synergistic chemotherapy. *ACS Applied Materials & Interfaces* 9:3487-3496.

Zipperer, Alexander, Konnerth, Martin C., Laux, Claudia, Berscheid, Anne, Janek, Daniela, Weidenmaier, Christopher, Burian, Marc, Schilling, Nadine A., Slavetinsky, Christoph, Marschal, Matthias, Willmann,

Matthias, Kalbacher, Hubert, Schittek, Birgit, Brotz-Oesterhelt, Heike, Grond, Stephanie, Peschel, Andreas, and Krismer, Bernhard 2016. Human commensals producing a novel antibiotic impair pathogen colonization. *Nature* 535:511-516.

BIOGRAPHICAL SKETCH

Jovana Vuković

Affiliation: Research Associate

Education:

2016. Doctor of Technical Sciences (PhD), University of Belgrade, Faculty of Technology and Metallurgy, Belgrade, Serbia

2008. Master of Engineering (MSc), University of Belgrade, Faculty of Technology and Metallurgy, Belgrade, Serbia

Business Address:

Innovation Center of Faculty of Technology and Metallurgy, University of Belgrade, Karnegijeva 4, 11000 Belgrade, Serbia

Research Experience:

International scientific project:

1. Intelligent Scaffolds as a Tool for Advanced Tissue Regeneration (Serbia-Slovenia-Switzerland, SCOPES – Swiss National Science Foundation, IZ73ZO_152327)

Projects funded by the competent Ministry of Education, Science and Technological Development of the Republic of Serbia:

1. Development of hydrogel-based nanocomposites for applications in reconstructive surgery, MNTR 19027, Belgrade (2008-2010),
2. The regeneration of skeletal tissue assisted by biomaterials as tissue matrices-*in vivo* and *in vitro* studies, MNTR 145072, Belgrade (2006-2010)
3. Chemical and structural design of nanomaterials for application in medicine and tissue engineering, MNTR 172026, Belgrade (2011-2016),
4. Synthesis and characterization of new functional polymers and polymer nanocomposites, MNTR 172062, Belgrade (2011-2016)

Professional Experience:

Employed at Innovation Center of Faculty of Technology and Metallurgy

Professional Appointments:

- Research associate 2017-present
- Research assistant 2011-2017
- Research trainee 2010-2011

Publications from the Last 3 Years:

1. Jovana S. Vuković, Aleksandra A. Perić-Grujić, Dragana S. Mitić-Ćulafić, Biljana Dj. Božić Nedeljković, Simonida Lj. Tomić, Antibacterial activity of pH-sensitive silver(I)/poly(2-hydroxyethyl acrylate/itaconic acid) hydrogels, Macromolecular Research (2019) DOI: https://doi.org/10.1007/s13233-020-8050-z, IF(2018)=1.758, ISSN 2092-7673.

Simonida Lj. Tomić

Affiliation: Faculty of Technology and Metallurgy, University of Belgrade, Belgrade, Serbia

Education: BSc, MSc, and Ph.D. at Faculty of Technology and Metallurgy, University of Belgrade, Belgrade, Serbia

Business Address: Faculty of Technology and Metallurgy, University of Belgrade, Karnegijeva 4, 11000 Belgrade, Serbia

Research and Professional Experience: Dr. Tomić completed her Ph.D. in Chemical Engineering in 2006 from Organic Chemical Technology Department at Faculty of Technology and Metallurgy, University of Belgrade (FTM-UB). She is currently the leader of an independent research group at FTM-UB dedicated to the synthesis and characterization of materials based on natural and synthetic polymers focused on the development of advanced, smart polymeric biomaterials and controlled drug delivery release systems (CDDRS) for applications in medicine and pharmacy. He has 40 peer-reviewed publications in high impact international journals, 2 book chapters and 3 chapters in monographs of national importance, 8 papers presented at international significance events printed in its integrity, 47 papers presented at international significance conferences printed in the form of abstracts, 4 papers presented at national importance assemblies printed in its integrity and 26 papers presented at national importance events. Dr. Tomić has published a prominent monograph of national importance. So far, she has been a mentor of 4 doctoral theses, 40 diploma works, and 5 master theses. Dr. Tomić's papers have been cited 809 times, of which 638 without autocitation. Dr. Tomić was the leader of an international project. Participated or participates in 3 international and 8 national projects. Dr. Tomić has invited reviewer in 48 international journals.

International scientific project:

1. Intelligent Scaffolds as a Tool for Advanced Tissue Regeneration (Serbia-Slovenia-Switzerland, SCOPES – Swiss National Science Foundation, IZ73ZO_152327)

Projects funded by the competent Ministry of Education, Science and Technological Development of the Republic of Serbia:

1. Development of hydrogel-based nanocomposites for applications in reconstructive surgery, MNTR 19027, Belgrade (2008-2010),
2. The regeneration of skeletal tissue assisted by biomaterials as tissue matrices-*in vivo* and *in vitro* studies, MNTR 145072, Belgrade (2006-2010)
3. Chemical and structural design of nanomaterials for application in medicine and tissue engineering, MNTR 172026, Belgrade (2011-2016),
4. Synthesis and characterization of new functional polymers and polymer nanocomposites, MNTR 172062, Belgrade (2011-2016)

Professional Appointments: University career-from 1995 to the present (assistant trainee, assistant, assistant professor, associate professor, and full professor).

Publications from the Last 3 Years:

1. Babić M. M., Tomić S. Lj. 2020. Semi-interpenetrating Networks Based on (Meth)acrylate, Itaconic Acid, and Poly(vinyl Pyrrolidone) Hydrogels for Biomedical Applications. In *Interpenetrating Polymer Network: Biomedical Applications,* edited by Sougata Jana, and Subrata Jana, 263-288, Springer Singapore.
2. Krezović B. D., Miljković M. G., Stojanović S. T., Najman S. J., Filipović J. M., Tomić S. Lj. 2017. Structural, thermal, mechanical, swelling, drug release, antibacterial and cytotoxic properties of

P(HEA/IA)/PVP semi-IPN hydrogels. *Chemical Engineering Research and Design* 121:368-80.

3. Filipović V. V., Bozić Nedeljković B. Dj., Vukomanović M., Tomić S. Lj. 2018. Biocompatible and degradable scaffolds based on 2-hydroxyethyl methacrylate, gelatin and poly(beta amino ester) crosslinkers, Polymer Testing 68:270-78.
4. Filipović V. V., Babić M. M., Gođevac D., Pavić A., Nikodinović-Runić J., Tomić S. Lj. 2019. *In Vitro* and *In Vivo* Biocompatibility of Novel Zwitterionic Poly(Beta Amino)Ester Hydrogels Based on Diacrylate and Glycine for Site-Specific Controlled Drug Release. *Macromolecular Chemistry and Physics* 220:1900188-
5. Babić M. M., Božić B. Đ., Božić B. Đ., Ušćumlić G. S., Tomić S. Lj. 2018. The innovative combined microwave-assisted and photo-polymerization technique for synthesis of the novel degradable hydroxyethyl (meth)acrylate/gelatin based scaffolds. *Materials Letters* 213:236-40.
6. Vuković J. S., Perić-Grujić A. A., Mitić-Ćulafić D. S., Božić Nedeljković B. Dj., Tomić S. Lj. 2020. Antibacterial Activity of pH-Sensitive Silver(I)/poly(2-hydroxyethyl Acrylate/itaconic acid) Hydrogels. *Macromolecular Research* 28:382-89.

In: An Introduction to Antibacterial Properties ISBN: 978-1-53618-305-4
Editor: Nicholas Paquette

Chapter 4

ANTIBACTERIAL DRUG RESISTANCE: THE CAUSES AND THE STRATEGIES TO OVERCOME THE DRUG RESISTANCE

Farha Naaz[1], Ritika Srivastava[2], Vishal K. Singh[1] and Ramendra K. Singh[1]
[1]Bioorganic Research Laboratory, Department of Chemistry
University of Allahabad, Prayagraj, India
[2]Department of Chemical Sciences, Indian Institute of Science
Education and Research Berhampur, Odisha, India

ABSTRACT

Antibiotics, addressed as the 'wonder drugs', are used for various therapeutic purposes for decades and a concomitant problem is the emergence of drug resistance, which is a challenging issue before the scientific community and the pharmaceutical industry.

Understanding the various aspects of the use of antibiotics, like targets of antibiotics, viz. cell wall synthesis, protein synthesis, nucleic acid synthesis and folate synthesis, the antibiotics used as inhibitors and their mechanisms of action, i.e., prevention of access to drug targets, changes in the structure and protection of antibiotic targets and the direct modification or inactivation of antibiotics, the emergence and causes of antibiotic resistance and strategies to overcome the antibiotic resistance are the focal points discussed in this chapter.

The antibiotic resistance crisis has been attributed to overuse and abuse of antibiotics, excessive prescribing, and slackness in new drug development by the pharmaceutical industry due to reduced economic incentives and daunting regulatory requirements. Designing novel analogs, drug combinations, innovative therapeutic approaches, and developing antibiotics with new mechanisms are considered to be the most effective strategies to combat the inevitable phenomena of drug resistance.

Keywords: antibiotics, antibiotic resistance, drug targets, mechanism of action, MIC

INTRODUCTION

Antibiotics, discovered long ago in the golden era (1940-1970), are compounds that target crucial bacterial functions or development processes, mainly aiming to interfere with or destroy the bacterial cell wall, the cell membrane, or crucial bacterial enzymes involved in nucleic acid and protein synthesis [1]. These "miracle drugs" proved to be very useful as therapeutic agents in the treatment of various bacterial infections, and thus inculcated confidence in human beings that infectious diseases can be controlled and prohibited [2].

Initially, the antibiotics were exceptionally useful but unfortunately, microorganisms hastily evolved and developed resistance against antimicrobial drugs leading to the standard treatments ineffective and the infection persisted and this resulted in turning some of these antibiotics outdated. Thus, the emergence of antimicrobial resistance appeared as a concomitant problem, which resulted in a serious challenge for the treatment of microbial infections. As a sequel to this, an alarming necessity for novel antibiotics emerged that rekindled the field of antibiotics and microbial research [3].

It was the use and misuse of antimicrobial drugs that accelerated the emergence of drug-resistant strains. Since the discovery of penicillin in 1928 by Fleming, microbial antibiotics have completely revolutionized the antibacterial therapy. Indeed, penicillin became the main therapeutic option for infectious diseases. However, during the 1940s widespread use of this antibiotic resulted in the emergence of new strains of microbes capable of destroying the drug and negating its effects. In the year 1947, a penicillin-resistant bacterial strain *Staphylococcus aureus* was discovered [4-5]. In 1959, the pharmaceutical industry developed methicillin, capable of avoiding penicillinase - the enzyme which degraded penicillin rings. Methicillin reached the market in 1960, however, *Staphylococcus aureus* strains resistant to methicillin appeared just one year later, proving how fast the bacteria evolved and became resistant [6-7]. The craving for a solution to this problem resulted in the discovery of some new antibiotics, viz. ampicillin, cephalosporins, vancomycin, and levofloxacin, before 2000 [8-9]. The first vancomycin-resistant bacteria (VRE, vancomycin-resistant enterococcus) were detected approximately after 30 years of medical use [10-11]. Additional options, like the use of linezolid and daptomycin, were exposed to quick needle resistance as well [12]. In 2017, a novel and durable vancomycin derivative was synthesized with exceptional features to circumvent the emergence of resistance (Figure 1).

Owing to the discovery break all through the last decades for novel antibiotic chemotherapy in the pharmaceutical field and to the incidence of bacterial strains resistant to the present antibiotics, public health is running out of treatment options in dealing with infectious diseases. In response to this emerging crisis, universal organizations such as the WHO has urged the scientific society to explore novel approaches to fight antibiotic resistance.

Enormous research efforts are focused on producing superior versions of existing molecules for developing new antibiotics as they are often less strenuous and prove better options than screening new compounds. However, some of the molecules thus produced, with only a little better activity, are simply "me-too" versions of the standard drugs. Hence, molecules developed through rational and methodical processes that are less vulnerable to resistance mechanisms are the need of the hour [14].

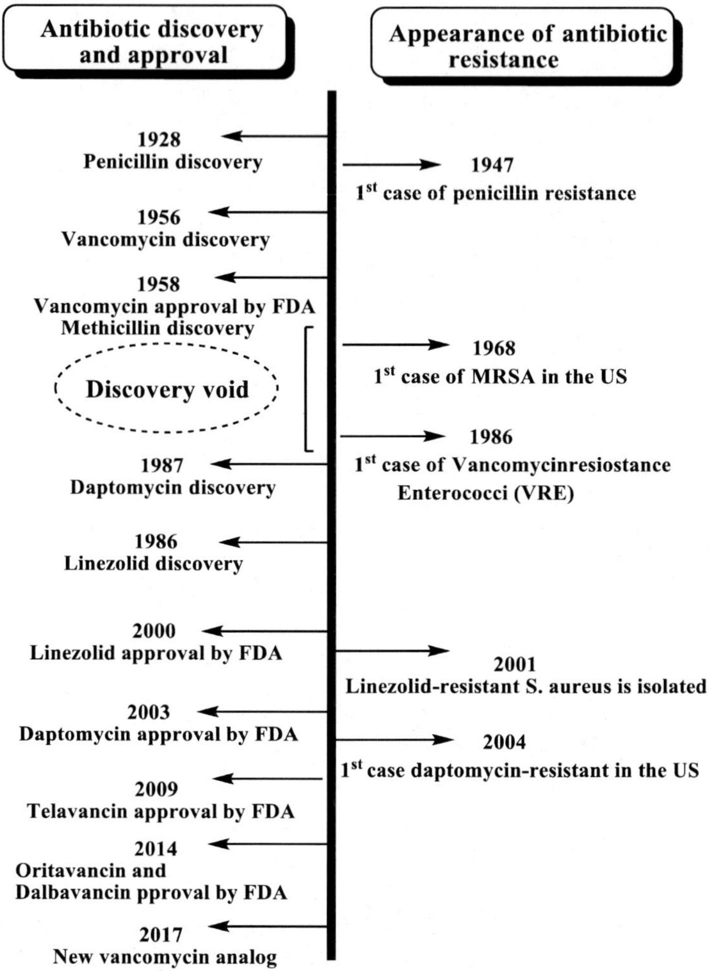

Figure 1. Timeline presentation of the discovery of antibiotics and the appearance of resistance

One of the vital phases of antibiotic drug discovery is the search for novel antibacterial targets. Alternatives to the most exploited antibacterial targets such as cell wall synthesis or protein synthesis are urgently needed to combat the drug-resistance problem. This chapter intends to focus on the mechanism of action and resistance development in commonly used antimicrobials and the strategies to overcome the resistance problem.

Antibacterial Targets Based on Mechanism of Action

Mechanistic and systemic knowledge of bacterial resistance provides much better opportunities to tackle this crisis by addressing the root of the problem rather than merely creating additional options of confusion in the way of drug development. Diverse classes of antibiotics efficient for treating particular types of bacteria are available nowadays. Antibiotics may be given orally, intravenously, or by intramuscular injection, depending on the type and severity of bacterial disease and other factors. The mechanism of resistance development and different targets of antibiotics are illustrated in Figure 2.

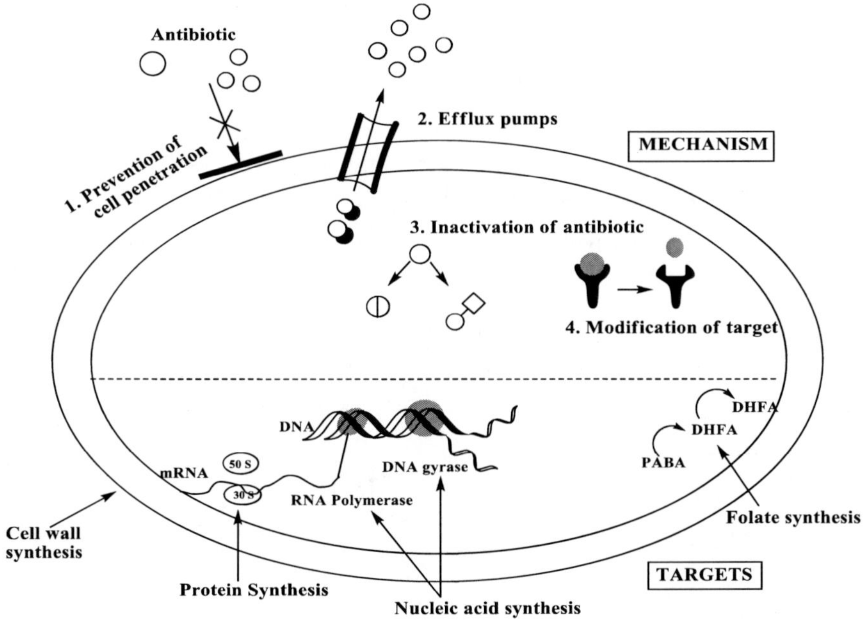

Figure 2. The mechanism of resistance development and main targets of antibiotics

Inhibitors of Cell Wall Synthesis

The cell wall that surrounds the bacterial cells is composed of peptidoglycans consisting of long sugar polymers and short peptides. The peptidoglycan layer is formed by cross-linking of the glycan strands and the peptide chains [15]. In the presence of penicillin-binding proteins (PBPs), the D-alanyl-alanine portion of the peptide chain is cross-linked by glycine residues [16]. This cross-linkage strengthens the bacterial cell wall. A large number of antibiotics target the peptidoglycan synthesis or its precursors, which leads to cell death. β-Lactams and glycopeptides are classical examples of antibiotics targeting the cell wall.

β-lactams

β-lactam antibiotics and their numerous scaffolds are still extensively used worldwide. The key targets of the antibiotics are the PBPs. The mechanism of action of β-lactams involves covalent binding to transpeptidases, thus inhibiting cell wall synthesis [17]. Indeed, the β-lactam ring mimics the D-alanyl-D-alanine terminal sequence of a peptide chain that is usually bound by PBP. The PBP interacts with β-lactam rings and is not essential for new peptidoglycan synthesis. Therefore, the antibiotic functions as a competitive antagonist to the transpeptidase enzyme. The disruption of the peptidoglycan layer leads to the inactivation of the enzyme and ultimately to the lysis and death of bacterium [18].

Glycopeptides

All glycopeptides share a general method of action, which consists of binding to the C-terminal D-alanyl-D-alanine of peptidoglycan precursors. This was first revealed in the case of vancomycin, a hydrophilic rigid glycopeptides, in the late 1960s. This binding prevents the transglycosylation and transpeptidation steps in the cell wall synthesis, eventually resulting in bacterial cell death [19].

Inhibitors of Protein Synthesis

Bacterial protein synthesis is a main antibiotic target and has been used for successful structure-based antibiotic resistance interventions [20-21]. Biosynthesis of proteins is catalyzed by the ribosome and cytoplasmic factors that transform mRNA into the respective polypeptide chain. The 70S bacterial ribosome consists of two subunits of ribonucleoprotein - the 30S and 50S subunits [22]. Antimicrobials hinder the protein biosynthesis by attacking the bacterial ribosome 30S (includes 16S rRNA) or 50S (includes 23S and 5S rRNAs) subunit [23-24].

Inhibitors of 30S Subunit

Aminoglycosides

The main target of the action of aminoglycosides (AGs) is the bacterial ribosome. The penetration of the cytoplasmic membrane requires an energy-dependent active bacterial transport mechanism, which needs oxygen and an active motive force of proton. AGs function in aerobic environments for these reasons and have low action against anaerobic bacteria. Such AGs have synergy with those antibiotics that inhibit cell wall synthesis (such as β-lactam and glycopeptides). AGs interact with the 16S rRNA, a subunit of 30S, near the 'A' site through hydrogen bonds and trigger mRNA translation to be misread and prematurely terminated [25-26].

Tetracyclines

Tetracyclines, recognized by four highly oxygenated fused rings, are broad-spectrum antibiotics with very few side effects. Tetracyclines, such as tetracycline, chlortetracycline, doxycycline, orminocycline, act on the retained sequences of the 30S ribosomal subunit 16S r-RNA to prevent t-RNA from binding to the 'A' site [27-28].

Inhibitors of 50S Subunit

Oxazolidinones

Linezolid and eperezolid are recently approved nontoxic members of a novel class of synthetic antibiotics. Oxazolidinones interfere with protein synthesis at several stages: (i) hinder the protein synthesis by binding to 23S rRNA of the 50S subunit at the peptidyl transferase center and (ii) influence the binding and orientation of the initiator-tRNA and prevent binding on the 'A' site, thereby disrupting the translation sequence [29-30].

Macrolides

Macrolides are bacteriostatic like many other classes of antibiotics and influence the early stage of protein synthesis, specifically translocation, by targeting the conserved sequences of the peptidyl transferase core of the ribosomal subunit 23S rRNA of 50S. This causes incomplete peptide chains to be removed prematurely [31].

Inhibitors of DNA Replication

Quinolones

Quinolones and fluoroquinolones are chemotherapeutic bactericidal antibiotics, eradicating bacteria by blocking the DNA replication pathway. Antibiotics of this class work by inhibiting one or more of a group of enzymes called topoisomerase, essentially required for supercoiling, replication, and separation of bacterial DNA strands. DNA gyrase is a topoisomerase II that catalyzes the negative supercoiling of the circular DNA found in bacteria. Topoisomerase IV, on the other hand, is involved in the relaxation of the supercoiled circular DNA, enabling the partition of the interwoven daughter chromosomes at the end of bacterial DNA replication. If an antibiotic binds to a bacterial enzyme, it may change the enzyme's activating site and prevent it from reacting with its substratum [32-33].

Table 1. Antibiotics and their respective targets of action

S.No.	Antibiotic Targets	Class	Example	Structure
1	Cell wall synthesis	β-lactam Glycopeptides Lantibiotics Fosfomycin	Sublactam	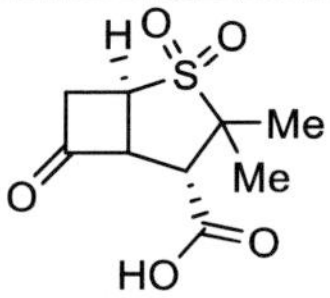
2	Protein Synthesis (50S)	Oxazolidinones Macrolides Thiopeptides Streptogramins Lincosamides Chloramphenicol	Eperezolid	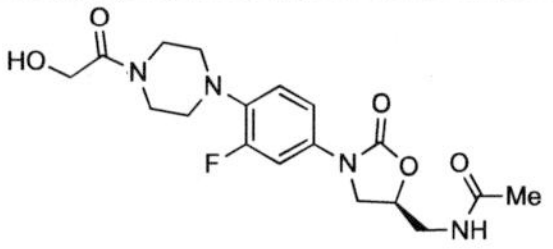
3	Protein Synthesis (30S)	Tetracycline Aminoglycosides	Chlortetracycline	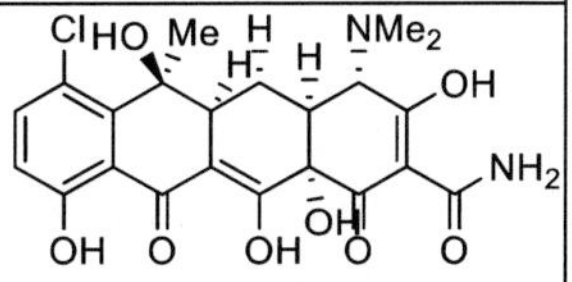
4	RNA polymerase	Rifamycins Actinomycins	Rifamycin B	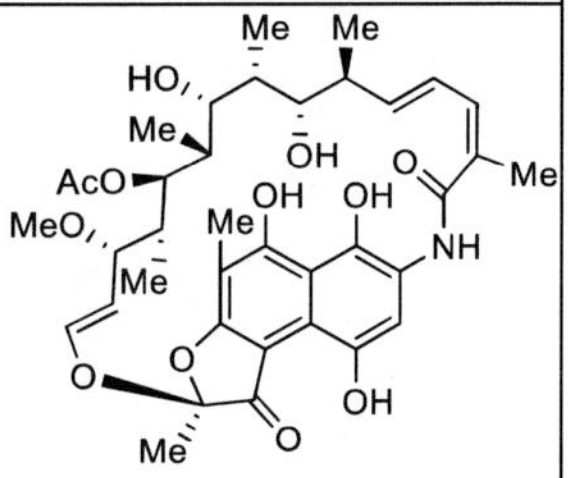
5	DNA gyrase	Fluoroquinolones Aminocoumarins	Ciprofloxacin	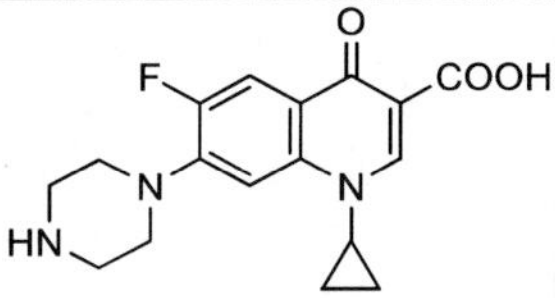
6	Folate mechanism	Sulfonamide Trimethoprim	Sulfamethoxazole	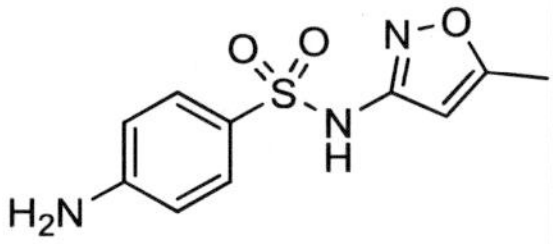

Inhibitors of Folic Acid Synthesis

Sulfonamides and Trimethoprim

Sulfonamides are bacteriostatic, broad-spectrum anti-infective agents. They are structural analogs of para-aminobenzoic acid (PABA) and competitively inhibit a bacterial enzyme, dihydropteroate synthetase, which is responsible for introducing PABA into the immediate precursor of folic acid, dihydrofolic acid. It inhibits the dihydrofolic acid synthesis and reduces the amount of metabolically active tetrahydrofolic acid, a cofactor for purine, thymidine, and DNA synthesis. At one step in the folate biosynthesis, trimethoprim and sulfonamides exert their effect in continuation. Hence, a combination of sulpha drugs and trimethoprim acting at different stages in the same biosynthetic pathway demonstrates synergy and reduces the resistance mutation rate [34].

A tabulated summary of these antibiotics along with their respective targets of action, different classes, examples, and structure are presented in Table 1.

MECHANISM OF ANTIBIOTIC RESISTANCE

The term "resistance" refers to the bacteria's ability to survive a particular antibiotic treatment. Some bacterial species are naturally immune to a given antibiotic group while others have acquired resistance. Acquired resistance means that only certain strains of a given species are immune to an antibiotic, but not the entire population. This resistance may arise owing to a random mutation in the chromosomal DNA or may occur extra-chromosomally, such as when plasmids or transposons are exchanged by bacteria. Some common examples of resistance mechanisms include modification/inactivation of the antibiotic itself, changes in the permeability of the external membrane, the presence of efflux pumps, and changes in the target bacterial site, as illustrated in Figure 2 [35-36].

Changes in External Membrane Permeability

As compared to gram-positive, gram-negative bacteria are generally less permeable to certain antibiotics since their outer membrane forms a barrier to permeability. Antibacterial agents can be delivered inside a cell by diffusion through porins and the bilayer, and by self-uptake. The porin channels are present in the outer membrane (OM) of gram-negative bacteria. The small hydrophilic molecules (β-lactams and quinolones) can traverse the OM solely through porins. Reducing the OM's permeability and limiting antibiotic entry into the bacterial cell is therefore accomplished by replacing porins with more-selective channels. The reduction in porin channels contributes to a diminution in the entry of β-lactam antibiotics and fluoroquinolone into the cell, resulting thereby in resistance to these antibiotic groups. Resistance acquired for all classes of antibiotics in *Pseudomonas aeruginosa* is due to low OM permeability [37-38].

Efflux Pumps

Membrane proteins are called efflux pumps, which export antibiotics from the cell and sustain their low intracellular concentrations. When the antimicrobials penetrate the cell at the same speed, the efflux mechanism pumps them out again before they hit their target. These pumps are found in the cytoplasmic membrane, in contrast to the porins found in OM. Antibiotics of all groups, except polymyxin, are susceptible to efflux system activation [39]. Efflux pumps may be antibiotic-resistant also. Many of them are multi-drug carriers capable of circulating abroad a variety of unidentified antibiotics - macrolides, tetracyclines, and fluoroquinolones and thereby leading significantly to multidrug-resistant species [40].

Modification or Complete Replacement of the Target Site

A common mechanism of resistance is natural variations or an acquired shift in antimicrobial target sites, which lowers or even prevents drug binding affinity. Changes in the target site frequently result from random chromosome mutation of a bacterial gene. Since antibiotic interaction with a target molecule is typically quite specific, any light alteration of the target molecule may have a significant effect on antibiotic binding.

- Alteration in the 30S subunit or 50S subunit

Modifications of these ribosome subunits result in resistance to antibiotics like macrolides, tetracycline, chloramphenicol, and aminoglycosides and this impairs the synthesis of proteins. Aminoglycosides bind to the ribosomal subunit 30S while chloramphenicol, macrolides, lincosamides and streptogram B bind to the ribosomal subunit 50S to inhibit protein synthesis [41].

- Alteration in penicillin-binding protein (PBP)

PBP alteration is a predominant mechanism of resistance to gram-positive bacteria, while the development of β-lactamases is a mechanism for resistance to gram-negative bacteria. Mutation of PBP results in a decreased affinity towards β-lactam antibiotics. This pathway makes *Enterococcus faecium* immune to ampicillin and *Streptococcus pneumonia* immune to penicillin. Similarly, the resistance to methicillin and oxacillinin *Staphylococcus aureus* is associated with the incorporation of a mobile genetic factor-"staphylococcal cassette chromosome mec" into the chromosome of *S. aureus* bearing the gene resistance mec A [42]. Mec A gene encodes the protein PBP2a, a new penicillin-binding protein used to modify a natural staphylococcal PBP. The PBP2a exhibits a high tolerance to β-lactam antibiotics. Methicillin-resistant *S. aureus* strains can be cross-resistant to β-lactam antibiotics, streptomycin, and tetracycline and more often to erythromycin [43].

- Modified precursors of cell wall

Glycopeptides, e.g., vancomycin or teicoplanin, can prevent cell wall synthesis in gram-positive bacteria by binding them to D-alanyl-D-alanine residues of peptidoglycan precursors. D-alanyl-alanine is modified to D-alanyl-lactate, and as a consequence the glycopeptides do not interact with peptidoglycans, resulting in intolerance to glycopeptides. *Enterococcus fecium* and *Enterococcus faecalis* strains have significant vancomycin and teicoplanin resistance (Van A-type resistance). Bacteria having resistance from Van B and Van C show resistance to vancomycin, and are also immune to teicoplanin [44].

- Mutated DNA gyrase and topoisomerase IV develop resistance to fluoroquinolone (FQ)

The quinolones bind to the 'A' subunit of DNA gyrase. The resistance process includes the alteration of two enzymes: DNA gyrase (coded by genes gyr A and gyr B) and topoisomerase IV (coded by genes par C and par E) [45]. Mutations in genes gyr A and par C result in loss of replication and thus FQ cannot bind to these enzymes. Mechanisms of ribosomal defense that confer tetracycline resistance and mutations of RNA polymerase which confer rifampicin resistance are some classic examples.

Direct Modification or Inactivation of Antibiotics

Three major enzymes inactivate antibiotics, viz. (i) β-lactamases, (ii) enzymes that alter aminoglycosides, and (iii) acetyltransferases of chloramphenicol (AACs) [46].

Beta-lactamases

β-lactamases hydrolyze nearly all β-lactams with an ester and amide bonds, such as penicillins, cephalosporins, monobactams, and carbapenems. β-Lactamases are usually dominant enzymes and are reported in many phylogenetic families as follows:

- *Class A β-lactamases*: Two commonly occurring Class A β-lactamases observed in Enterobacteriaceae members are called TEM-1 and SHV-1. These are penicillinase with minimal to no cephalosporin action [47]. They are progenitors of β-lactamases in the extended-spectrum (ESBL). ESBL are enzymes that have modified the substratum profile due to amino acid substitution which helps most cephalosporins to hydrolyze. ESBL are immune to penicillins, cephalosporins of the third generation but vulnerable to methoxy-cephalosporins.
- *Class B β-lactamases*: The metallo-β-lactamases that require cofactors for their activity, such as zinc or heavy metals fall under this category and their function is inhibited by chelating agents. Such enzymes are immune to clavulanate, sulbactam, aztreonam, and carbapenem inactivation [48].
- *Class C β-lactamases*: These are also known as cephalosporinases. These are produced by all gram-negative bacteria except *Salmonella* and *Klebsiella.* Class C hydrolyzes cephalosporins including extended-spectrum cephalosporins. In comparison to class A β-lactamases, these have large cavities, and as a result, they can bind the bulky extended-spectrum penicillins. The Amp C β-lactamases are an example of this kind. This enzyme class is resistant to all β-lactams, except for carbapenems [49]. They are not inhibited by clavulanate.

Aminoglycoside Modifying Enzymes (AGEs)

Specific enzymes neutralize aminoglycosides, viz. phosphoryl-transferases, nucleotidyl-transferases, or adenylyl-transferases, and AACs. These aminoglycoside-modifying enzymes (AMEs) minimize the affinity of a transformed molecule, inhibit binding to the 30S ribosomal subunit [50], and have extended aminoglycoside and fluoroquinolone range resistance [51]. AMEs are recognized in *S. aureus*, *E. faecalis*, and *S. pneumonia* strains.

Chloramphenicol-acetyl-transferases

A few gram-positive and gram-negative bacteria and some of *Haemophilus influenza* strains are resistant to chloramphenicol, and they have an enzyme chloramphenicol transacetylase that acetylates hydroxyl groups of chloramphenicol. Modified chloramphenicol is unable to bind to a ribosomal 50S subunit properly.

The resistance mechanism of various reported antibiotics is described in Table 2.

Table 2. Resistance category and resistance mechanism of different antibiotics

S. No.	Antibiotic class	Type of resistance: Resistance mechanism
1	Aminoglycoside	Decreased uptake: Change in the outer membrane Enzymatic modification: AGEs
2	Beta-lactams	Altered PBP: PBP 2a Enzymatic degradation: Penicillinases which are classified as per ambler classification
3	Glycopeptides	Altered target: D-alanyl-alanine is changed to D-alanyl-D-lactate
4	Macrolides	Altered target: Methylation of ribosomal active site with reduced binding Efflux pumps: Mef type pump
5	Oxazolidinones	Altered target: Mutation leading to reduce binding to the active site
6	Quinolones	Altered target: Mutation leading to reduce binding to the active site(s) Efflux: Membrane transporters
7	Tetracyclines	Efflux: New membrane transporters Altered target: Production of proteins that bind to the ribosome and alter the conformation of the active site
8	Sulfa drugs	Altered target: Mutation of genes encoding DHPS

DHPS=Dihydropteroate synthase; PBP=Penicillin binding protein; AGEs=Aminoglycoside enzyme

Causes of Antibiotic Resistance

Overuse and Misuse of Antibiotics

The problem of antibiotic resistance in bacteria was caused by excessive and overuse of antibiotics. Epidemiological studies have shown a strong association between antibiotic use and the appearance and spread of resistant strains of bacteria. In bacteria, genes can be inherited from relatives or acquired from non-relatives on mobile genetic elements such as plasmids, thus allowing for the transfer of antibiotic resistance between different bacterial species. Resistance can also appear through mutations spontaneously. Antibiotics remove drug-sensitive competitors and leave resistant bacteria behind as a result of natural selection to reproduce. Despite warnings about overuse, antibiotics are overprescribed globally [52].

Inappropriate Diagnosis and Prescription

Antibiotics administered inappropriately often lead to the development of resistant bacteria. Research has shown that in 30 to 50 percent of cases, medication indication, choice of drug, or length of antibiotic therapy was false. The concentration of subinhibitory and subtherapeutic antibiotics may foster antibiotic resistance by promoting genetic modifications, such as improvements in gene expression and mutagenesis. Changes in antibiotic-induced gene expression can improve virulence, whereas increased mutagenesis can promote and spread antibiotic resistance [53].

Extensive Agricultural Application

It is claimed that treating livestock with antimicrobials increases the optimal health of the livestock, delivering better yields and a higher quality product [54]. The transfer of resistant bacteria to humans by farm animals was first noted more than 35 years ago when high rates of antibiotic resistance were found in the intestinal flora of both farm animals and farmers. More recently, molecular detection methods have demonstrated that resistant bacteria in farm animals reach consumers through meat products. This occurs through the following sequence of events: i) antibiotic use in food-producing animals kills or suppresses susceptible bacteria, allowing antibiotic-resistant bacteria to thrive; ii) resistant bacteria are transmitted to humans through the food supply; and, iii) these bacteria can cause infections in humans that may lead to adverse health consequences.

The use of antibiotics in agriculture also impacts environmental microbiome [55].

Less Accessibility to New Antibiotics

The pharmaceutical industry's production of new antibiotics regularly, a technique that had been successful in battling resistant bacteria in the past, was effectively blocked by economic and regulatory barriers. Antibiotic production is no longer considered an economically prudent venture for the pharmaceutical industry [56], as antibiotics are used for fairly brief periods and are mostly curative; hence the production of new antibiotics is not as lucrative as medications to cure medical conditions such as diabetes, mental illnesses, asthma, or gastro-oesophageal reflux. Since medications are more effective for medical diseases, pharmaceutical firms tend to invest in such fields.

Regulatory Barriers

Even those organizations that are positive about exploring the development of new antibiotics, consider regulatory clearance an obstacle [57]. Changes in the U.S. requirements for execution of clinical trials, during the last two decades by the Food and Drug Administration (FDA), have found clinical trials of antibiotics extremely difficult. The difficulties noted in seeking regulatory approval include complexity, lack of clarification, discrepancies in standards for clinical trials among countries, changes in regulatory and licensing laws, and inadequate communication networks. Additional new regulatory strategies are required to ensure that antibiotic medicines continue to evolve and become available.

Strategies to Overcome Resistance

Alternative therapeutic approaches have been developed to address the growing evolution of antimicrobial resistance with the common goal of reducing the number of antibiotics used and maintaining the existing groups of antibiotics for future clinical use. Some alternative solutions for solving the antibiotic resistance issue are discussed below:

Design of Novel Derivatives of Known Drugs

One strategy is to design or grow analogs of drugs that are currently in therapeutic use and have anti-resistant organisms function. For example, modified erythromycins were shown to overcome the resistance induced by the mechanism of ribosome methylation of macrolide-lincosamide-streptogram B (MLS) [58]. Many lipophilic tetracycline analogs, such as aminocycline, are less vulnerable to pathways of resistance to tetracycline obtained from plasmid glycopeptides. Alternatively, certain antibiotics obtained from natural sources are known but have not been produced historically due to lack of scope, possible toxicity, or other factors. For example, 16-membered macrolides, such as spiramycin, are active against inducible MLS-resistant strains [59] because they do not induce resistance to MLS.

Considering the crisis of antibiotic resistance and in surge for novel broad-spectrum antibacterial agents, we have synthesized and subjected some novel *N*-heterocyclic sulfonamide derivatives to *in vitro* antibacterial testing against four panels of gram +ve bacterial strains, viz. *Bacillus cereus*, *Staphylococcus aureus*, *Bacillus subtilis*, *Staphylococcus pyrogen*s, and gram –ve bacterial strains, viz. *Escherichia coli*, *Pseudomonas aeruginosa*, *Pseudomonas vulgaris, Pseudomonas mirabilis* for measuring the minimum inhibitory concentration (MIC, μg/mL). A detailed analysis of *in vitro* activity, validated by the findings of *in silico* studies, explicitly suggested that these active compounds could be identified as lead inhibitors of folic acid synthesis. The results further suggested that a combination approach for the production of highly potent drug regimens and novel formulations could be more useful. Antibacterial activities of some sulfonamide derivatives, synthesized in our laboratory as potential antibacterial agents [60] are summarized in Table 3.

The antibacterial results were quite rewarding and all four compounds exhibited significantly higher antibacterial activity and excellent inhibition pattern against the tested strains. It is noteworthy to mention here that some of the compounds were even much more potent than the reference used - sulfamethoxazole. Therefore, the study further indicated the development of highly potent drug regimens using the combination approach (detailed in our previous publication [60]).

Table 3. *In vitro* antibacterial activity of some *N*-heterocyclic sulfonamide derivatives against gram +ve and gram –ve bacterial strains

S. No	Compounds	MIC(μg/mL)							
		Gram +ve strain				Gram –ve strain			
		A	B	C	D	E	F	G	H
1	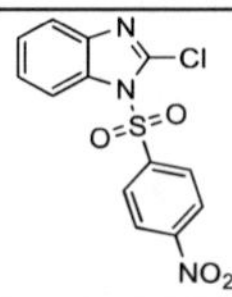	6.2	6.2	6.2	6.2	3.1	6.2	3.1	3.1
2	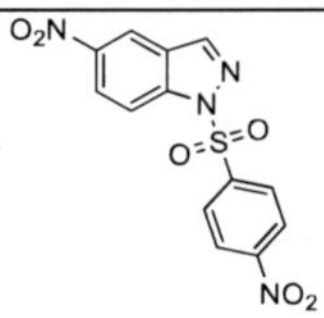	6.2	6.2	6.2	6.2	3.1	3.1	3.1	3.1
3	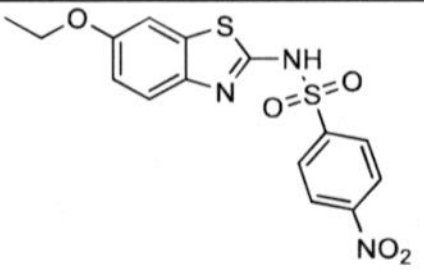	12.5	6.2	6.2	6.2	3.1	6.2	3.1	6.2
4	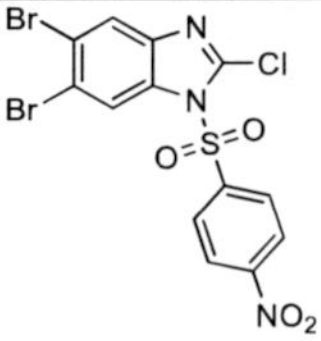	3.1	6.2	3.1	3.1	3.1	3.1	3.1	3.1
Ref.	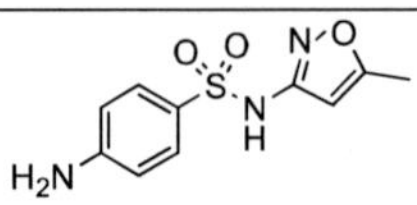	6.2	6.2	6.2	6.2	3.1	3.1	3.1	3.1

Ref=Sulfamethoxazole; A=*B.cerus*; B=*S. aureus*; C=*B.subtilis*; D=*S.pyrogenes*; E=*E.coli*; F=*P.aeruginosa*; G=*P.vulgaris*; H=*P.mirabilis*; MIC=Minimum Inhibitory Concentration.

Combinations of Drugs

When describing the molecular resistance process, the designing of different therapeutic combinations based on the antagonism of the significant resistance determinants has been made feasible. For example, the advent of amoxicillin-clavulanate hybrid therapy [61] in which amoxicillin is a highly beneficial antibiotic (orally absorbed, wide range, non-toxic, safe for pediatric use, and bactericidal) coupled with clavulanic acid, 1-lactamase inhibitor responsible for the majority of amoxicillin resistance, has proved quite satisfactory. These formulations have met with considerable clinical success.

Novel Therapeutic Approaches

Sometimes clinical care needs to focus on combined therapy without the production and application of new or enhanced antibiotics to ensure the coverage of resistant species. Attention must be renewed on the production and use of the vaccines. In the near future, approval of guidelines for safe antibiotic use [62], with a focus on the use of narrow-spectrum drugs over more wide-spectrum agents is expected to minimize the selective resistance risk. The course of antimicrobial drug development should, in any case, prioritize the treatment of resistant bacteria, preferably by following the suitable modes of action that do not support resistance emergence.

New Agents with Novel Mechanisms

As reported in a previous research article [63], the discovery and development of new antibacterial drugs have generally taken a target-oriented approach. One approach to resolve resistance has been to study inhibitors of novel targets so that new chemical species that are not sensitive to known resistance mechanisms can be discovered.

Conclusion

The need to create new agents for treating bacterial infections that have become increasingly unresponsive to conventional antibacterial therapy is presently acknowledged all over the world. Over the last few decades, antibiotic resistance has increasingly developed as one of the 21st century's biggest public health challenges. In fact, infections that are untreatable because of the infecting organism's multidrug tolerance have become more prevalent in clinical settings. This devastating situation has been made worse by a shortage of antibiotic research and production. The "golden" antibiotic research era during the 1960s and 1970s needs to be revived in the light of the present needs as the pharmaceutical industries focused their attention on other fields that were more lucrative and rewarding, leaving the surge of resistance to rise unabated.

Thorough knowledge of the mechanisms by which bacteria become immune to antibiotics is of utmost importance for the discovery of new approaches to combat the risk of resistance. We need to introduce antibiotics, with the expectation that the microorganism will respond to them and that resistance (an evolutionary fact) will evolve. Developing alternative/synergistic antibiotic treatments is the necessity for present-day therapeutics as the emergence of the pre-antibiotic age threatens mankind. Efforts to develop antibiotics and to examine resistance mechanisms should, therefore, be persistent, robust, and steady. Coordinated actions are desperately required to impose new strategies, accelerate research initiatives, and take measures to address the crisis. This is likely to be a long-haul "war" against living organisms with significant adaptability and survival capability.

REFERENCES

[1] Fernebro, J., (2011). Fighting bacterial infections-Future treatment options. *Drug Resist. Updat*. 14, 125-139.

[2] Demain, A. L. and Sanchez, S., (2009). Microbial drug discovery: 80 years of progress. *J. Antibiot*. 62, 5-16.

[3] Nisha, A., (2008). Antibiotic residues-a global health hazard. *Vet. World* 1, 375-377.

[4] Fleming, A., (1929). On the antibacterial action of cultures of a penicillium, with special reference to their use in the isolation of *B. influenzæ*. *Br. J. Exp. Pathol.*10, 226-236.

[5] Rammelkamp, C. H. and Maxon, T., (1992). Resistance of *Staphylococcus aureus* to the action of penicillin. *Proc. Soc. Exp. Biol. Med.* 51, 386-389.

[6] Enright, M. C., Robinson, D. A., Randle, G., Feil, E. J., Grundmann, H. and Spratt, B. G, (2002). The evolutionary history of methicillin-resistant *Staphylococcus aureus* (MRSA). *Proc. Natl. Acad. Sci.* USA, 99, 7687-7692.

[7] Waness, A., (2010). Revisiting methicillin-resistant *Staphylococcus aureus* infections. *J. Glob. Infect. Dis.* 2, 49-56.

[8] Lemke, T., (2008). *Foye's Principles of Medicinal Chemistry*, 6th ed.; Lippincott Williams & Wilkins: Philadelphia, PA, USA.

[9] Levine, D. P., (2006). Vancomycin: A history. *Clin. Infect. Dis.* 42 (Suppl.1), S5-S12.

[10] Murray, B. E., (2000). Vancomycin-resistant Enterococcal infections. *N. Engl. J. Med.* 342, 710-721.

[11] Mc Kessar, S. J., Berry, A. M., Bell, J. M., Turnidge, J. D. and Paton, J. C., (2000). Genetic characterization of van G, a novel vancomycin resistance locus of Enterococcus faecalis. *Antimicrob. Agents Chemother.* 44, 3224-3228.

[12] Brickner, S. J., Barbachyn, M. R., Hutchinson, D. K. and Manninen, P. R., (2008). Linezolid (ZYVOX), the first member of a completely new class of antibacterial agents for treatment of serious gram-positive infections. *J. Med. Chem.* 51, 1981-1990.

[13] *WHO Antimicrobial resistance: global report on surveillance 2014*, ISBN 9789241564748.

[14] Kardos, N. and Demain, A. L. (2011). Penicillin: The medicine with the greatest impact on therapeutic outcomes. *Appl. Microbiol. Biotechnol.* 92, 677-687.

[15] Kahne, D., Leimkuhler, C., Lu, W. and Walsh, C. (2005). Glycopeptide and lipoglycopeptide antibiotics. *Chem Rev.* 105, 425-48.

[16] Reynolds, P., E. (1989). Structure, biochemistry and mechanism of action of glycopeptides antibiotics. *Eur. J. Clin. Microbiol. Infect. Dis.* 8, 943-50.

[17] Fisher, J., F., Meroueh, S., O. and Mobashery, S. (2005). *Chem. Rev.*105, 395-424.

[18] Kishida, H., S., Unzai, Roper, D., I., Lloyd, A., Park, S., Y. and Tame, J., R., H. (2006). *Biochemistry.* 45,783-792.

[19] Reynolds, P., E. (1989). Structure, biochemistry and mechanism of action of glycopeptide antibiotics. *Eur. J. Clin. Microbiol. Infect. Dis.* 8, 943-950.

[20] Poehlsgaard, J. and Douthwaite, S. (2005). The bacterial ribosome as a target for antibiotics. *Nat. Rev. Microbiol.* 3, 870-881.

[21] Wilson, D., N. (2009). The A-Z of bacterial translation inhibitors. *Crit. Rev. Biochem. Mol. Biol.* 44, 393-433.

[22] Moore, P. (2012). How Should We Think About the Ribosome? *B. Annu. Rev. Biophys.* 41, 1-19.

[23] Vannuffel, P. and Cocito, C. (1996). Mechanism of action of streptogramins and macrolides. *Drugs* 51 Suppl1, 20-30.8.

[24] Johnston, N., J., Mukhtar, T., A. and Wright, G., D. (2002). Streptogramin antibiotics: Mode of action and resistance. *Curr. Drug Targets* 3, 335-344.

[25] Magnet,S. and Blanchard, J., S. (2005). The Kinetic Mechanism of AAC (3)-IV Aminoglycoside Acetyltransferase from *Escherichia coli. Chem. Rev.* 105, 477-498.

[26] Rodnina, M., V. and Wintermeyer, W. (2001). Ribosome fidelity: tRNA discrimination, proof reading and induced fit. *Trends Biochem. Sci.* 26, 124-130.

[27] Chopra, I. and Roberts, M. (2001). Tetracycline Antibiotics: Mode of Action, Applications, Molecular Biology, and Epidemiology of Bacterial Resistance. *Microbiol. Mol. Biol. Rev.* 65, 232-260.

[28] Blanchard, S., C., Gonzalez, R., L., Kim, H., D., Chu, S. and Puglisi, J., D. (2004). tRNA selection and kinetic proof reading in translation. *Nat. Struct. Mol. Biol.* 11, 1008-1014.

[29] Lambert, P., A. (2005). Bacterial resistance to antibiotics: Modified target sites. *Adv Drug Deliv Rev.* 57, 1471-1485.

[30] Bozdogan, B. and Appelbaum, P., C. (2004). Oxazolidinones: Activity, mode of action, and mechanism of resistance. *Int J Antimicrob Agents.* 23, 113-119.

[31] Mankin, A., S. (2008). Macrolide Myths. *Curr. Opin. Microbiol.* 11, 414-421.

[32] Yoneyama, H. and Katsumata, R. (2006). Antibiotic resistance in bacteria and its future for novel antibiotic development. *Biosci Biotechnol Biochem.* 70, 1060-1075.

[33] Higgins, P., G., Fluit, A., C. and Schmitz, F., J. (2003). Fluoroquinolones: Structure and target sites. *Curr Drug Targets* 4, 181-190.

[34] Argyropoulou, I., Geronikaki, A., Vicini, P. and Zanib, F. (2009). Synthesis and biological evaluation of sulfonamide thiazole and benzothiazole derivatives as antimicrobial agents. *Arkivoc.* 6, 89-112.

[35] Munita, J., M. and Arias, C., A. (2016). Mechanisms of Antibiotic Resistance. *Microbiol. Spectr.* 4.

[36] Pages, J., M., James, C., E. and Winterhalter, M. (2008). The porin and the permeating antibiotic: A selective diffusion barrier in gram-negative bacteria. *Nat. Rev. Microbiol.* 6, 893-903.

[37] Kojima, S. and Nikaido, H. (2013). Permeation rates of penicillins indicate that *Escherichia coli* porins function principally as non specific channels. *Proc. Natl Acad. Sci.* USA110, E2629-E2634.

[38] Tamber, S. and Hancock, R., E. (2003). On the mechanism of solute uptake in Pseudomonas. *Front. Biosci.* 8, s472-s483.

[39] Dzidic, S., Suskovic, J. and Kos, B. (2008). Antibiotic resistance mechanisms in bacteria: Biochemical and genetic aspects. *Food Technol Biotechnol.* 46, 11-21.

[40] Lambert, P., A. (2002). Mechanisms of antibiotic resistance in *Pseudomonas aeruginosa. J R SocMed.* 95, Suppl41,22-6.

[41] Tenover, F., C. (2006). Mechanisms of antimicrobial resistance in bacteria. *Am J Med.* 1196 Suppl1, S3-10.

[42] Alekshun, M., N. and Levy, S., B. (2007). Molecular mechanisms of antibacterial multidrug resistance. *Cell* 128, 1037-1050.

[43] Hiramatsu, K., Cui, L., Kuroda, M. and Ito, T. (2001). The emergence and evolution of methicillin-resistant *Staphylococcus aureus. Trends Microbiol.* 9, 486-493.

[44] Giedraitiene, A., Vitkauskiene, A., Naginiene, R. and Pavilonis, A. (2011). Antibiotic resistance mechanisms of clinically important bacteria. *Medicina (Kaunas)* 47, 137-146.

[45] Kim, Y., H., Cha, C., J. and Cerniglia, C., E. (2002). Purification and characterization of an erythromycin esterase from an erythromycin-resistant Pseudomonas sp. FEMS *Microbiol Lett.* 210, 239-244.

[46] Dockrell, H., M., Goering, R., V., Roitt, I., Wakelin, D. and Zuckerman, M. (2004). Attacking the enemy: Antimicrobial agents and chemotherapy. In: Mims C., Dockrell H., M., Goering, R., V., Roitt,I., Wakelin, D., Zuckerman, M., editors. *Medical Microbiology.* 473-507.

[47] Rice, L., B., Sahm, D. and Bonomo, R. (2003). Mechanisms of resistance to antibacterial agents. In: Murray, P., R., editor. *Manual of Clinical Microbiology.* 8thed. Washington, D., C.: ASM Press.1084-7.

[48] Bonnet, R. (2004). Growing group of extended-spectrum beta-lactamases: The CTX-Menzymes. *Antimicrob. Agents Chemother.* 48, 1-14.

[49] Queenan, A., M. and Bush, K. (2007). Carbapenemases: the versatile β-lactamases. *Clin. Microbiol. Rev.* 20, 440-458.

[50] Strateva, T. and Yordanov, D. (2009). Pseudomonas aeruginosa–A phenomenon of bacterial resistance. *J Med Microbiol.* 58, 1133-1148.

[51] Maurice, F., Broutin, I., Podglajen, I., Benas, P., Collatz, E. and Dardel, F. (2008). Enzyme structural plasticity and the emergence of broad-spectrum antibiotic resistance. *EMBORep.*9, 344-349.

[52] The antibiotic alarm (2013). *Nature* 495 (7440),141.

[53] Viswanathan, V., K. (2014). Off-label abuse of antibiotics by bacteria. *Gut Microbes* 5, 3-4.

[54] Bartlett, J., G., Gilbert, D., N. and Spellberg, B. (2013). Seven ways to preserve the miracle of antibiotics. *Clin Infect Dis*. 56, 1445-1450.

[55] Michael, C., A., Dominey-Howes, D. and Labbate, M. (2014). The antibiotic resistance crisis: causes, consequences, and management. *Front Public Health* 2, 145.

[56] Piddock, L., J. (2012). The crisis of no new antibiotics-what is the way forward? *Lancet Infect Dis*. 12, 249-253.

[57] Gould, I., M. and Bal, A., M. (2013). New antibiotic agents in the pipeline and how they can overcome microbial resistance. *Virulence* 4, 185-191.

[58] Goldman, R., C., and Kadam, S., K. (1989). Binding of novel macrolide structures to macrolide-lincosamide-streptogramin B-resistant ribosomes inhibits protein synthesis and bacterial growth. *Antimicrob. Agents Chemother*. 24, 851-862.

[59] Hardy, D., J., Hensey, D., M., Beyer, J., M., Vojtko, C., McDonals, E., J. and Fernandes, P., B. (1988). Comparative *in vitro* activities of new 14-, 15-, and 16-membered macrolides. *Antimicrob. Agents Chemother*. 32, 1710-1719.

[60] Naaz, F., Srivastava, R., Singh, A., Singh, N., Verma, R., Singh, V., K. and Singh, R., K. (2018). Molecular modeling, synthesis, antibacterial and cytotoxicity evaluation of sulfonamide derivatives of benzimidazole, indazole, benzothiazole and thiazole; *Bioorganic & Medicinal Chemistry* 26, 3414-3428.

[61] Bush, K. (1988). Lactamase inhibitors from laboratory to clinic. *Clin. Microbiol. Rev*. 1, 109-123.

[62] Marr, J., J., Moffet, H., L. and Kunin, C., M. (1988). Guidelines for improving the use of antimicrobial agents in hospitals: a statement by the Infectious Disease Society of America. *J. Infect. Dis*. 157, 869-876.

[63] Silver, L. and Bostian, K. (1990). Screening of natural products for antimicrobial agents. *Eur. J. Clin. Microbiol. Infect. Dis*. 9, 455-461.

BIOGRAPHICAL SKETCHES

Ramendra K. Singh

Affiliation: University of Allahabad

Education: MSc, PhD

Business Address: Bioorganic Research Laboratory
Department of Chemistry
Faculty of Science, University of Allahabad
Prayagraj–211002, INDIA
Tel:+91.532.2461005(Lab)/2461236(O)/2250391(res);
Mob. 9450304598
Email: rksinghsrk@gmail.com

Research and Professional Experience:

Dr. Ramendra K. Singh, Professor of Chemistry, is Director of Bioorganic Research Laboratory and Controller of Examination in the University of Allahabad, India. He has more than 30 years of experience in the field of nucleic acids research and about two decades in the field of antiretroviral research with wide range of training in various national and international institutions. His research interests also include developing natural compound curcumin as anti-cancer agent and fluorescently labeled nucleosides/oligonucleotides as markers in drug delivery. He is member of several national bodies, editor to different international journals and referee to more than two dozen national/international journals.

Professional Appointments: Professor (2012)

Associate Professor (2009), Lecturer (1997)

Honors: Young Scientist's Award (ISCA), India;
XVIIUBMB Fellowship, India;
Jawaharlal Nehru Visiting Fellowship, India;
UNESCO Fellowship, Japan;
Post-Doc Fellowship, Japan;
INSA International Exchange Fellowship, Poland
Fulbright Fellowship, USA

Publications from the Last 3 Years:

Research articles:

1. Vishal K. Singh, Ritika Srivastava, Parth Sarthi Sen Gupta, Farha Naaz, Himani Chaurasia, Richa Mishra, Malay Kumar Ranaand and Ramendra K. Singh (2020). Anti-HIV potential of diaryl pyrimidine derivatives as non-nucleoside reverse transcriptase inhibitors: design, synthesis, docking, TOPKAT analysis and molecular dynamics simulations. *Journal of Biomolecular Structure and Dynamics,* doi: 10.1080/07391102.2020.1748111.
2. Yogesh K. Pandey, Anu Mishra, Pratibha Rai, Jaya Singh, Jagdamba Singh and Ramendra K. Singh (2020) DBU Catalysis: An Efficient Synthetic Strategy for 5,7-disubstituted-1,2,4-triazolo[1,5-a]pyrimidines. *Current Organic Synthesis*, 17(1), 73-80.
3. Madhu Yadav, Ritika Srivastava, Farha Naaz, Anuradha Singh, Rajesh Verma and Ramendra K. Singh (2019). *In silico* studies on new oxathiadizoles as potential anti-HIV agents, *Gene Reports*, 14, 87-93.
4. Madhu Yadav, Ritika Srivastava, Farha Naaz, and Ramendra K. Singh (2018). Synthesis, docking, ADMET prediction, cytotoxicity and antimicrobial activity of oxathiadizole derivatives, *Computational Biology and Chemistry*, 77, 226-239.
5. Ritika Srivastava, Sunil K. gupta, Farha Naaz, Anuradha Singh, Vishal Kumar Singh, Rajesh Verma, Nidhi Singh and Ramendra K.

Singh (2018). Synthesis, antibacterial activity, synergistic effect, cytotoxicity docking and molecular dynamics of benzimidazole analogues, *Computational Biology and Chemistry*, 76, 1-16.

6. Farha Naaz, Ritika Srivastava, Anuradha Singh, Nidhi Singh, Rajesh Verma, Vishal Kumar Singh and Ramendra K. Singh (2018). Molecular modeling, synthesis, antibacterial and cytotoxicity evaluation of sulfonamide derivatives of benzimidazole, indazole, benzothiazole and thiazole, *Bioorganic and Medicinal Chemistry*, 26, 3414-3428.
7. Singh, V. K., Singh, R. Verma and Ramendra K. Singh (2018) *In silico* studies on *N*-(pyridine-2-yl) thiobenzamides as NNRTIs against wild and mutant HIV-1 Strains. *Philippine Journal of Science*, 147(1), 37-46.
8. Anuradha Singh, Ritika Srivastava, and Ramendra K. Singh (2017). Design, synthesis, and antibacterial activities of novel heterocyclic aryl sulphonamide derivatives, *Interdisciplinary Sciences Computational Life Sciences*, doi 10.1007/s12539-016-0207-2.
9. M. Yadav, A. Singh and Ramendra K. Singh (2017). Docking studies on novel bis phenyl benzimidazoles (BPBIs) as non-nucleoside inhibitors of HIV-1 reverse transcriptase. *Indian Journal of Chemistry (B)*, 56B, 714-723.
10. Nidhi Singh, Ritika Srivastava, Anuradha Singh and Ramendra K. Singh (2016). Synthesis & Photophysical studies on naphthalimide derived fluorophores as markers in drug delivery, *Journal of Fluorescence*, 26, 1431-1438.
11. Anuradha Singh, Madhu Yadav, Ritika Srivastava, Nidhi Singh, Rajinder Kaur, Satish K. Gupta and Ramendra K. Singh (2016). Design and anti-HIV activity of aryl sulphonamides as non-nucleoside reverse transcriptase inhibitors, *Medicinal Chemistry Research*, 25, 2842-2859.
12. Garima Kumari and Ramendra K. Singh (2016). Molecular modeling, synthesis and anti-HIV activity of novel isoindoline dione analogues as potent NNRTIs. *Chemical Biology & Drug Design*, 87, 200-212.

Chapters in books

1. Ritika Srivastava, Farha Naaz and Ramendra K. Singh (2019). "Role of *in silico* and *invitro* cytotoxicity evaluation in development of potential drugs" in *Advances in Medicine and Biology, Volume 144* (Leon V. Berhardt-Editor) pp11-44, Nova Science Publishers, 2019, USA, ISBN: 978-1-53615-842-7.
2. Ritika Srivastava, Farha Naaz, Madhu Yadav, Vishal Kumar Singh and Ramendra K. Singh (2019). "Prodrugs: An advanced approach for the drug development to enhance the therapeutic efficacy" in *Advances in Medicine and Biology, Volume 147* (Leon V. Berhardt-Editor) pp 75-112, Nova Science Publishers, 2019, USA, ISBN: 978-1-53616-062-8.
3. Vishal K. Singh, Richa Mishra, Himani Chaurasia, Ritika Srivastava, Vivek K. Chaturvedi, and Ramendra K. Singh (2018). "Natural Products as Anti-HIV Agents and their Role in HIV-Associated Neurocognitive Disorders (HAND) in "*Innovations in Agricultural and Biological Engineering*". Apple Academic Press, USA. In press.
4. Ramendra K. Singh and Anuradha Singh (2017). "Human Immunodeficiency Virus Reverse Transcriptase (HIV-RT): Structural Implications for Drug Development" in "*Recent advancements in Pharmaceutical, Nutritional and Industrial Enzymology*" (S. L. Bharati and P. K. Chaurasia–eds), IGI Global, In Press.

Farha Naaz

Affiliation: University of Allahabad

Education: M.Sc., PhD

Business Address: C/o Prof Ramendra K Singh
Bioorganic Research Laboratory, Department of Chemistry,
University of Allahabad, Prayagraj–211002,INDIA
Tel: +91-532-2461005(lab); Mob.9670283540
Email: farhanaaz88aualld@gmail.com

Research and Professional Experience: Research experience of more than 5years.

Professional Appointments: D. Phil. Candidate

Honors: Honored with gold medal award during graduation studies and recipient of CSIR-SRF.

Publications from the Last 3 Years:

Research articles:

1. Vishal K. Singh, Ritika Srivastava, Parth Sarthi Sen Gupta, Farha Naaz, Himani Chaurasia, Richa Mishra, Malay Kumar Rana & Ramendra K. Singh (2020). Anti-HIV potentia l of diaryl pyrimidine derivatives as non-nucleoside reverse transcriptase inhibitors: design, synthesis, docking, TOPKAT analysis and molecular dynamics simulations. *Journal of Biomolecular Structure and Dynamics,* doi: 10.1080/07391102.2020.1748111.
2. Madhu Yadav, Ritika Srivastava, Farha Naaz, Anuradha Singh, Rajesh Verma and Ramendra K. Singh (2019). *In silico* studies on new oxathiadizoles as potential anti-HIV agents, *Gene Reports*, 14, 87-93.
3. Madhu Yadav, Ritika Srivastava, Farha Naaz, and Ramendra K. Singh (2018). Synthesis, docking, ADMET prediction, cytotoxicity and antimicrobial activity of oxathiadizole derivatives, *Computational Biology and Chemistry*, 77, 226-239.

4. Ritika Srivastava, Sunil K. gupta, Farha Naaz, Anuradha Singh, Vishal Kumar Singh, Rajesh Verma, Nidhi Singh and Ramendra K. Singh (2018). Synthesis, antibacterial activity, synergistic effect, cytotoxicity docking and molecular dynamics of benzimidazole analogues, *Computational Biology and Chemistry*, 76, 1-16.
5. Farha Naaz, Ritika Srivastava, Anuradha Singh, Nidhi Singh, Rajesh Verma, Vishal Kumar Singh and Ramendra K. Singh (2018). Molecular modeling, synthesis, antibacterial and cytotoxicity evaluation of sulfonamide derivatives of benzimidazole, indazole, benzothiazole and thiazole, *Bioorganic and Medicinal Chemistry*, 26, 3414-3428.

Chapters in books

1. Ritika Srivastava, Farha Naaz and Ramendra K. Singh (2019). "Role of *in silico* and *in vitro* cytotoxicity evaluation in development of potential drugs" in *Advances in Medicine and Biology, Volume 144* (Leon V. Berhardt-Editor) pp 11-44, Nova Science Publishers, 2019, USA, ISBN: 978-1-53615-842-7.
2. Ritika Srivastava, Farha Naaz, Madhu Yadav, Vishal Kumar Singh and Ramendra K. Singh (2019). "Prodrugs: An advanced approach for the drug development to enhance the therapeutic efficacy" in *Advances in Medicine and Biology, Volume 147* (Leon V. Berhardt -Editor) pp 75-112, Nova Science Publishers, 2019,USA, ISBN: 978-1-53616-062-8.

Ritika Srivastava

Affiliation: University of Allahabad

Education: M.Sc., PhD

Business Address: c/o Prof Ramendra K Singh
Bioorganic Research Laboratory, Department of Chemistry,
University of Allahabad, Prayagraj–211002, INDIA
Tel: +91-532-2461005(lab); Mob.08303191415/7309270137
E-mail: ritika.ritu.au@gmail.com

Research and Professional Experience: Research experience of more than 5 years (2013-2019).

Professional Appointments: Post-Doc Fellow, IISER, Berhampur, Odisha

Honors: Swarna Jayanti Fellowship (2018) by National Academy of Sciences, India.

Publications from the Last 3 Years:

Research articles:

1. Vishal K. Singh, Ritika Srivastava, Parth Sarthi Sen Gupta, Farha Naaz, Himani Chaurasia, Richa Mishra, Malay Kumar Rana & Ramendra K. Singh (2020). Anti-HIV potential of diaryl pyrimidine derivatives as non-nucleoside reverse transcriptase inhibitors: design, synthesis, docking, TOPKAT analysis and molecular dynamics simulations. *Journal of Biomolecular Structure and Dynamics,* doi: 10.1080/07391102.2020.1748111.
2. Madhu Yadav, Ritika Srivastava, Farha Naaz, Anuradha Singh, Rajesh Verma and Ramendra K. Singh (2019). *In silico* studies on new oxathiadizoles as potential anti-HIV agents, *Gene Reports*, 14, 87-93.
3. Madhu Yadav, Ritika Srivastava, Farha Naaz, and Ramendra K. Singh (2018). Synthesis, docking, ADMET prediction, cytotoxicity and antimicrobial activity of oxathiadizole derivatives, *Computational Biology and Chemistry*, 77, 226-239.

4. Ritika Srivastava, Sunil K. gupta, Farha Naaz, Anuradha Singh, Vishal Kumar Singh, Rajesh Verma, Nidhi Singh and Ramendra K. Singh (2018). Synthesis, antibacterial activity, synergistic effect, cytotoxicity docking and molecular dynamics of benzimidazole analogues, *Computational Biology and Chemistry*, 76, 1-16.
5. Farha Naaz, Ritika Srivastava, Anuradha Singh, Nidhi Singh, Rajesh Verma, Vishal Kumar Singh and Ramendra K. Singh (2018). Molecular modeling, synthesis, antibacterial and cytotoxicity evaluation of sulfonamide derivatives of benzimidazole, indazole, benzothiazole and thiazole, *Bioorganic and Medicinal Chemistry*, 26, 3414-3428.
6. Anuradha Singh, Ritika Srivastava, and Ramendra K. Singh (2017). Design, synthesis, and antibacterial activities of novel heterocyclic aryl sulphonamide derivatives, *Interdisciplinary Sciences Computational Life Sciences*, doi: 10.1007/s12539-016-0207-2.
7. Nidhi Singh, Ritika Srivastava, Anuradha Singh and Ramendra K. Singh (2016). Synthesis & Photophysical studies on naphthalimide derived fluorophores as markers in drug delivery, *Journal of Fluorescence*, 26, 1431-1438.
8. Anuradha Singh, Madhu Yadav, Ritika Srivastava, Nidhi Singh, Rajinder Kaur, Satish K. Gupta and Ramendra K. Singh (2016). Design and anti-HIV activity of aryl sulphonamides as non-nucleoside reverse transcriptase inhibitors, *Medicinal Chemistry Research*, 25, 2842-2859.

Chapters in books

1. Ritika Srivastava, Farha Naaz and Ramendra K. Singh (2019). "Role of *in silico* and *in vitro* cytotoxicity evaluation in development of potential drugs" in *Advances in Medicine and Biology, Volume 144* (Leon V. Berhardt-Editor) pp 11-44, Nova Science Publishers, 2019, USA, ISBN: 978-1-53615-842-7.
2. Ritika Srivastava, Farha Naaz, Madhu Yadav, Vishal Kumar Singh and Ramendra K. Singh (2019). "Prodrugs: An advanced approach

for the drug development to enhance the therapeutic efficacy" in *Advances in Medicine and Biology, Volume 147* (Leon V. Berhardt-Editor) pp 75-112, Nova Science Publishers, 2019, USA, ISBN: 978-1-53616-062-8.

Vishal Kumar Singh

Affiliation: University of Allahabad

Education: M.Sc., PhD

Business Address: C/o Prof Ramendra K Singh
Bioorganic Research Laboratory, Department of Chemistry,
University of Allahabad, Prayagraj-211002, INDIA
Tel: +91-532-2461005(lab); Mob.7398146209
E-mail: vishalkumarsingh922@gmail.com

Research and Professional Experience: Research experience of more than 4 years.

Professional Appointments: D. Phil. Candidate

Honors: Recipient of UGC-JRF.

Publications from the Last 3 Years:

Research articles:

1. Vishal K. Singh, Ritika Srivastava, Parth Sarthi Sen Gupta, Farha Naaz, Himani Chaurasia, Richa Mishra, Malay Kumar Rana & Ramendra K. Singh (2020). Anti-HIV potential of diaryl pyrimidine derivatives as non-nucleoside reverse transcriptase inhibitors: design, synthesis, docking, TOPKAT analysis and molecular

dynamics simulations. *Journal of Biomolecular Structure and Dynamics,* doi: 10.1080/07391102.2020.1748111.

2. Ritika Srivastava, Sunil K. gupta, Farha Naaz, Anuradha Singh, Vishal Kumar Singh, Rajesh Verma, Nidhi Singh and Ramendra K. Singh (2018). Synthesis, antibacterial activity, synergistic effect, cytotoxicity docking and molecular dynamics of benzimidazole analogues, *Computational Biology and Chemistry*, 76, 1-16.
3. Farha Naaz, Ritika Srivastava, Anuradha Singh, Nidhi Singh, Rajesh Verma, Vishal Kumar Singh and Ramendra K. Singh (2018). Molecular modeling, synthesis, antibacterial and cytotoxicity evaluation of sulfonamide derivatives of benzimidazole, indazole, benzothiazole and thiazole, *Bioorganic and Medicinal Chemistry*, 26, 3414-3428.
4. Anuradha Singh, Vishal Kumar Singh, Rajesh Verma and Ramendra K Singh (2018). *In silico* studies on *N*-(pyridine-2-yl) thiobenzamides as NNRTIs against wild and mutant HIV-1 strains, *Philippine Journal of Science*, 147, 37-46.

Chapters in books

1. Ritika Srivastava, Farha Naaz, Madhu Yadav, Vishal Kumar Singh and Ramendra K. Singh (2019). "Prodrugs: An advanced approach for the drug development to enhance the therapeutic efficacy" in *Advances in Medicine and Biology, Volume 147* (Leon V. Berhardt-Editor) pp 75-112, Nova Science Publishers, 2019, USA, ISBN: 978-1-53616-062-8.

In: An Introduction to Antibacterial Properties ISBN: 978-1-53618-305-4
Editor: Nicholas Paquette

Chapter 5

ANTIBIOTIC RESISTANCE BREAKERS: STRATEGIES TO COMBAT THE ANTIBACTERIAL DRUG RESISTANCE

Anuradha Singh
Department of Chemistry, Sadanlal Sanwaldas Khanna Girls' Degree College (A Constituent College of the University of Allahabad)
Prayagraj, India

ABSTRACT

Antibacterial resistance is currently a global challenge since the number of resistant strains against multiple antibiotics continuously increasing, and there is an urgent need to develop novel strategies to overcome this problem. The outer protective membrane and vital overexpressed efflux pumps are some key factors responsible for intrinsic resistance in Gram-negative bacteria.

The antibiotic drug discovery process comprising chemical modification of existing antibiotic scaffolds has been proven very successful to expand potency, spectrum, and bypass resistance pathways. However, the emergence of drug-resistant microbial strain severely compromised the effectiveness of currently available commercial antibiotics. The antimicrobial resistance is the result of evolution and an unavoidable condition, which leads to the conclusion that developing an alternative perspective for treatment options is vital. Though, many strategies may be employed to minimize the impact and emergence of resistance, among them combination therapies are one of such effective strategies specially adjuvants that are chemically active moieties with no antibiotic action. Antimicrobial adjuvants acted as a blocker of antimicrobial resistance or booster of antimicrobial action. This approach not only suppresses the emergence of bacterial resistance but rejuvenate the antimicrobial activity of currently available commercial antibiotics cost-effectively.

Keywords: antibiotics, antibiotic resistance, mechanism of action, combination therapy, adjuvants

INTRODUCTION

One of the greatest triumphs of scientific research was the discovery of wonder drug penicillin and its antimicrobial action by Alexander Fleming in 1928. This great discovery started the era of antibiotics and revolutionized modern healthcare landscape after the Second World War. During this period, the importance of wonder antibiotics is embellished by the four Nobel Prizes in medicine for the discoveries of the sulphonamide prontosil (1939), penicillin (1945), streptomycin (1952), and artemisinin and avermectin (2015) [1, 2]. Antibiotics now have become an integral part of the modern healthcare practices, which not only have dramatically reduced the death toll of previously life-threatening infectious diseases and it enabled surgeries and other medical interventions as well. Thus, the discovery, commercialization and routine administration of wonder drugs against microbial infections revolutionized modern medicine and changed the therapeutic paradigm [3, 4].

Unfortunately, continuous use of antibiotics, self-medication, and hospital-acquired infection has induced the emergence of antimicrobial resistance (AMR), one of the three most important public health threats of the 21st century by WHO (World Health Organization) [5, 6]. The rapid global dissemination of AMR not only jeopardized public health advances but has also ensued interruption in the natural eco-system by establishing of resistant microbes. The ESKAPE (*Enterococcus faecium*, *Staphylococcus aureus, Klebsiella pneumoniae, Acinetobacter baumannii, Pseudomonas aeruginosa, and Enterobacter species*) bacteria associated with nosocomial infections worldwide are quite alarming. Further, four of six species in the ESKAPE bacterias are Gram-negative, which are particularly difficult to treat. ESKAPE pathogens are capable of "escaping" the biocidal action of antimicrobial agents, responsible for multidrug resistance (MDR), resulted in extended hospital stays, frequent doctor visits, psychological upsets, and case fatality are some issues associated with high health care costs. With every passing year, the overall number of antibiotics effective against ESKAPE is declining [7, 8].

The combination of antimicrobial resistance and a dearth of a robust antibiotic pipeline marks the onset of a possible severe worldwide health crisis, in both hospital and community settings, predisposing us toward a post-antibiotic era where common infections and minor injuries may result in significant morbidity and mortality. A recent report estimated that antibiotic resistance will cause around 300 million premature deaths by 2050, with a loss of up to $100 trillion to the global economy [9].

Though the rate of antibiotic consumption and resistance frequently arising, the dearth of a robust antibiotic pipeline is still a major challenge before the scientific community. The economic dysfunction within the pharmaceutical companies, long and difficult regulatory processes, lack of funding for a clinical trial are some key barriers that have impeded attempts to discover new antibiotics.

To curtail the emergence of AMR, the use of combinations therapy based on a cocktail of commercially available drugs emerges as the most viable strategy. Combinations therapy includes antibiotic-antibiotic combinations and the pairing of an antibiotic with a non-antibiotic molecule which also known as adjuvant or resistance breakers to target resistance mechanisms. The development of Antibiotic Resistance Breakers (ARBs) which potentiate the action of already existing antibiotics are the most innovative and attractive part of cocktail therapy. ARBs, when conjugate with old antibiotics, can revive the original potency against the resistant strain of both Gram-positive and Gram-negative bacteria.

In this chapter, the concept of combination therapy with special reference to antibiotic resistance breakers has been discussed in detail.

UNDERSTANDING OF ANTIMICROBIAL RESISTANCE

The comprehension of antimicrobial resistance and strategies adopted by bacteria to steer clear of antibiotics is necessary to understand the potentiate action of already existing antibiotics. The bacterial resistance to antimicrobial drugs is inbuilt of a bacterial evolutionary process which is facilitated by sophisticated mechanisms. In other words, antimicrobial resistance is ancient and it is the outcome of the interaction between microbes and their environment. Almost all the primary antibiotics are naturally-produced molecules developed during microbial metabolism in the form of bioactive chemicals. Thus co-resident bacteria which are "intrinsically" resistant to one or more antimicrobials have evolved to overcome their action to survive. It is noteworthy that in discussing the antimicrobial resistance mystery, intrinsic bacterial resistance is not the centre of the problem. The central point of discussion in clinical settings is preferably "acquired bacterial resistance" that was originally susceptible to the antimicrobial compound. The acquired resistance can be accomplished by either mutation in chromosomal genes or due acquisition of external genetic determinants of resistance, possibly obtained from intrinsically resistant organisms already present in the environment. Further, resistance to one antimicrobial class can usually be achieved through multiple biochemical pathways, and one bacterial cell may be capable of using a series of mechanisms of resistance to overcome an antibiotic effect [10-14]. Generally, the following three conditions are required to show antibiotic activity [15]:

a) The presence of the well-defined intact target in the bacterial cell.
b) To trigger the desired biological effect, sufficient quantity antibiotic is required at the proper target in the bacterial cell.
c) The antibiotic should be active and does not undergo any change within the cell.

The alteration of one of these conditions may hamper desired antibiotic action and leads bacterial resistance towards particular antibiotic. The defence strategies (resistance mechanisms are shown in Figure 1) adopted by bacteria to combat the effects of antibiotics are as follows [15-16]:

Restrict Access of the Antibiotic by Deactivating Outer Membrane Permeability

Bacteria restrict access by changing the entryways or limiting the number of entryways. Gram-positive and Gram-negative bacteria differ in the composition of their cell envelope. The cell envelope of Gram-positive bacteria consists of an inner plasma membrane surrounded by the cell wall, a thick layer of peptidoglycan, which comprises the outermost layer of the cell. The cell wall is permeable and typically does not restrict the diffusion of antibiotics into the cell. Gram-negative bacteria meanwhile possess a much thinner peptidoglycan layer that is surrounded by a second membrane comprised of a bilayer of phospholipids and lipopolysaccharide (LPS), known as the outer membrane, which is the outermost structure of the cell. The outer membrane of Gram-negative bacteria provides an extra layer of protection for the cell as compared to Gram-positive bacteria and plays a major role in preventing the diffusion of hydrophobic molecules, including many antibiotics, into the cell. As a result, these compounds can only enter the cell through selective porins, providing an intrinsic resistance of Gram-negative bacteria to many antibiotics, although they possess the intercellular targets of these drugs.

Get Rid of the Antibiotic by Activation of Efflux Pump

Bacteria get rid of antibiotics using pumps in their cell walls to remove antibiotic drugs that enter the cell. The production of complex bacterial machinery capable to the reduction of the antibiotic influx or increase of its efflux out of the cell and prevent it from reaching its target in sufficient quantity can also result in antimicrobial resistance. Tetracycline resistance is one of the classic examples of efflux-mediated resistance, where the Tet efflux pumps (belonging to the MFS family) extrude tetracyclines using proton exchange as the source of energy. Some *Pseudomonas aeruginosa* bacteria can produce pumps to get rid of several different important antibiotic drugs, including fluoroquinolones, beta-lactams, chloramphenicol, and trimethoprim. The major survival weapon of Gram-negative bacteria is its intrinsic resistance that renders them refractory to current antibiotics. This arises from highly effective protective barrier formed by the lipopolysaccharides (LPS) on their outer membrane (already discussed), and second, constitutively overexpressed efflux pumps preventing most molecules from permeating into the bacteria and reach their target.

Change or Destroy the Antibiotic by Enzymatic Inactivation

An existing cellular enzyme is modified to react with the antibiotic in such a way that it no longer affects the bacteria but deactivating processes degrade the antibiotic and reduces the affinity of the antibiotic for its target. Example: *Klebsiella pneumoniae* bacteria produce enzymes called carbapenemases, which break down carbapenem drugs and most other beta-lactam drugs. The most frequent biochemical reactions they catalyze include acetylation (aminoglycosides, chloramphenicol, streptogramins), phosphorylation (aminoglycosides, chloramphenicol), and adenylation (aminoglycosides, lincosamides).

Bypass the Effects of the Antibiotic

Bacteria can develop new cell processes that avoid using the antibiotic's target. Example: Some *Staphylococcus aureus* bacteria can bypass the drug effects of trimethoprim.

Change the Targets for the Antibiotic by Modification on the Drug Target Site

Many antibiotic drugs are designed to target specific vital pathway of a bacterium. Bacteria change the antibiotic's target so the drug can no longer fit and do its job because there is complementary interaction between antibiotic and target to show its desire activity. Hence, any slight structural change of the target may compromise the binding affinity of antibiotics significantly. For example, *E. coli* bacteria with the *mcr-1* gene can add a compound to the outside of the cell wall so that the drug colistin cannot latch onto it. The macrolides and tetracycline resistance are in this category.

Moreover, antibiotic-resistant bacteria may overexpress genes that encode molecular defence mechanisms such as efflux pumps or drug-inactivating enzymes. These resistance genes can disseminate to a different organism via horizontal gene transfer of mobile genetic elements such as plasmids, transposons, and integrons.

Thus, the need for novel antimicrobial drugs is a continuous one to counteract the development of bacterial resistance. There is a serious imbalance in the permanent race between bacterial evolution and the number of effective antibiotics. It is worth mentioning that during the late 1960s, the exploration of new antibiotics with novel mechanisms of action severely declined. The last novel class of antibiotic was discovered was in 1968. The subsequent antibiotics were modified versions of previously discovered antibiotic classes. The development of new antibacterial agents has decreased substantially in recent decades despite the current demand for new antimicrobial drugs; perhaps the most suitable reason is high cost and low profitability of pursuing new drug development within the pharmaceutical industry. The number of newly approved antibacterial agents has decreased over 40 years from 1983 to 2020 [17].

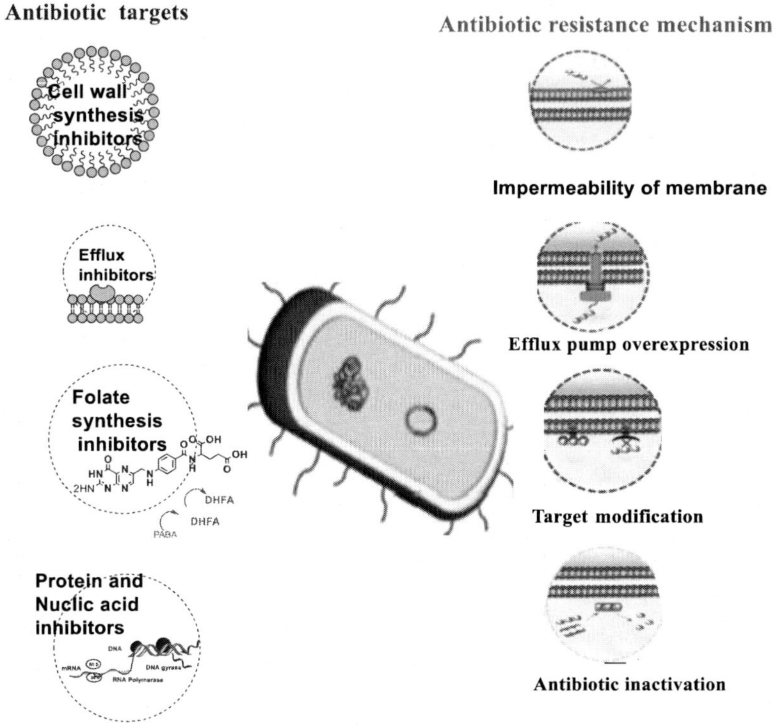

Figure 1. Antibiotic targets and bacterial resistance mechanisms.

Both conventional and non-conventional approaches are needed to address this pressing clinical problem. The contemporary researches are based on several potential avenues and multitude of a different perspective to revive the therapeutic arsenal, viz. the modification of existing drugs, the discovery of novel antibiotic templates, and the design of combination antibiotic adjuvants that don't kill bacteria, but only inhibit their infectivity, the use of antibodies or phages [18-19].

This chapter highlights the strategies on how to move forward in this bacterial resistance era via the development of combinations of antibiotics and adjuvants, which improve antibiotic performance by hitting multiple bacterial targets.

COMBINATION THERAPY TO COMBAT AMR

The monotherapy which defined by one drug–one target model has limited viability, and combination therapy is the norm in the treatment of many cancers, viral infections such as HIV, and tuberculosis for more than 3 decades [20]. The use of drug cocktails such as antibiotic/adjuvant combinations against AMR is an attractive alternative to the fuel the current arsenal of available antibiotics. This approach could be a possible solution to emerging AMR by using combinations of existing antibiotics and revival adjuvants, to potentiate the antibiotic against the resistant strain of interest [21-23]. The Key features of combination therapy have been shown in Figure 2.

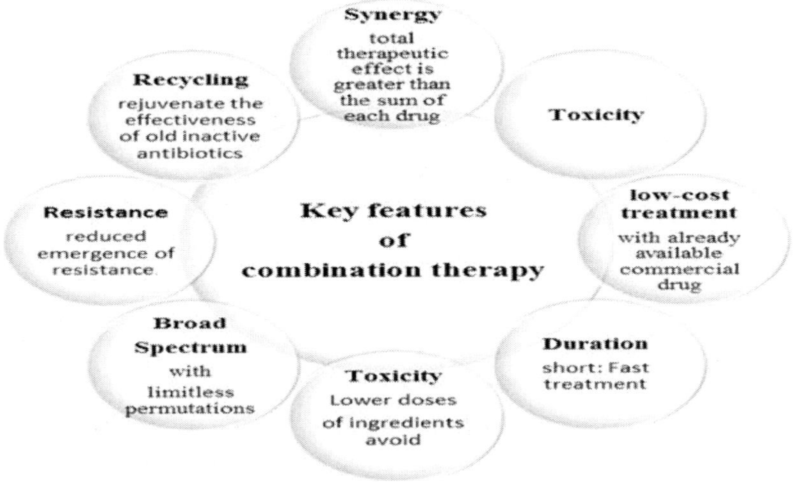

Figure 2. Key features of combination therapy.

In the combination of two or more drugs, if the total therapeutic effect is greater than the sum of each drug, there exists a synergistic effect in drug combinations [24]. For example, Zhou *et al.* reported the synergistic combinations of vancomycin and nitrofurantoin; vancomycin and trimethoprim against E. coli. vancomycin is a glycopeptide antibiotic that blocks peptidoglycan polymerization but it cannot penetrate the outer membrane of Gram-negative bacteria such as *E.coli* [25]. A large number of already available commercial antibiotics and their limitless synergistic permutations provide a vast potential for therapeutic gains. Moreover, this approach not only rejuvenates the effectiveness of old antibiotics against bacteria which have previously developed resistance against them but also lowers their resistant profile with a short duration of therapy. For example, monotherapy leads to resistance and disease relapse for tuberculosis, while, combination therapy provides reduced resistance profile with a short duration of therapy for the same. The fast and low-cost empirical combination therapy widens the coverage of clinically suspected bacterial infections of unknown cause. Thus, compared to the very expensive traditional monotherapy approach, combination therapy is a wise choice already used by Mother Nature. With added perks of lower resistance emergence, this strategy could enable us to keep pace with AMR emergence [26].

Classification of Synergistic Antibiotic Combinations

In combining bioactive compounds, is to achieve synergy, following combinations has been reported (Figure 3) [27]:

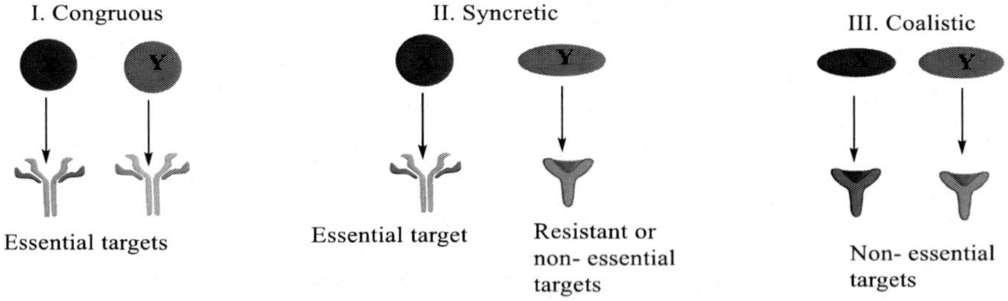

Figure 3. Classification of synergistic antibiotic combinations.

Congruous Combinations

Antibiotics are based on compounds that individually have cell growth inhibition activity towards the target organism. As shown in figure 3, in a congruous pair, two antibiotics (X and Y) that target distinct essential molecular processes can display synergy. Example: Following synergistic combinations offers broad-spectrum coverage of both Gram-positive and Gram-negative pathogens.

- Combination of penicillin with streptomycin for enterococcal infections
- Combination of rifampin–isoniazid– pyrazinamide in the treatment of tuberculosis.
- Formulated fixed-dose antibiotic combinations are: Co-trimoxazole (trimethoprim and sulfamethoxazole in a 1:5 weight/weight (w/w) ratio), which is sold under various trade names, including Septra and Bactrim. Synercid is a synergistic combination of streptogramin antibiotics comprising quinupristin and dalfopristin in a 3:7 w/w ratio. Topical agents such as bacitracin and polymyxin B (Polysporin; sometimes with the addition of gramicidin) and Neosporin, which combines neomycin, bacitracin and gramicidin are also commercially available antibiotic drug combinations.

The well-documented use of congruous antibiotic combinations to achieve broad-spectrum coverage in the case where the infective organism is unknown and where the need for rapid treatment is acute remains the best argument for an empirical combination of antibiotics. The disadvantage in such applications is the opportunity for unnecessary antibiotic exposure that fuels resistance in the patient and the health- care setting.

Syncretic Combinations

The syncretic combinations paradigm entails the use of bioactive adjuvants that augment the antibiotic efficacy of a primary antibiotic against drug-resistant pathogens. These compounds are also termed as "resistance breakers" or "antibiotic potentiators." The adjuvant has no antibacterial activity on its own but can either serve as antibiotic resistance breakers or antibiotic booster. An antibiotic (X) that targets an essential process and a non-antibiotic adjuvant (Y), the molecular target of which is a resistance element or a non- essential bacterial or host target as depicted in Figure 3. Antibiotic adjuvants further divided according to the scheme shown in Figure 4:

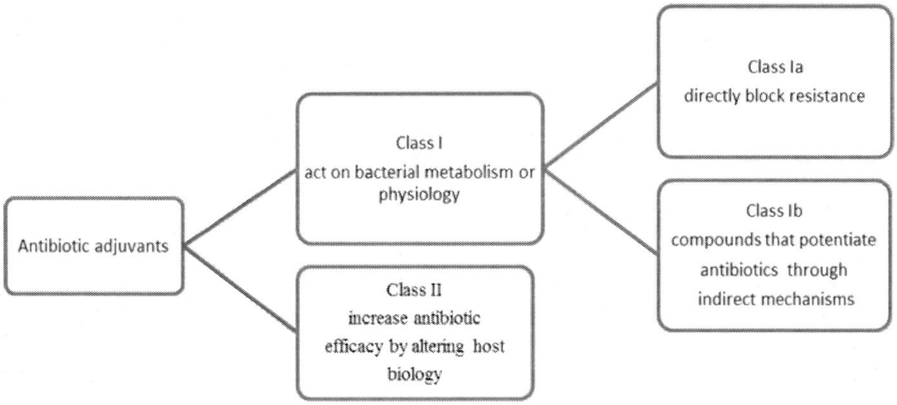

Figure 4. Classification of antibiotic adjuvants.

Class I Syncretic Adjuvants

These adjuvants act on bacterial metabolism or physiology. Class I adjuvants can further be differentiated based on mechanism of action. The compounds that directly block resistance are grouped under Class Ia, which can be further classified depending on the resistance mechanism which they oppose: adjuvant may be an efflux pump inhibitor (EPI) (to prevent the extrusion of drugs), a membrane permeabilizer (to increase the number of molecules that penetrate the membrane), or an enzyme inhibitor (to prevent the degradation of drugs before they reach their targets). The compounds that potentiate antibiotics through indirect mechanisms grouped under Class

Ib. Some representative examples of this class have been shown in Figure 5. Adjuvant effect of clavulanic acid on antibiotic synergy is as follow:

The combination of clavulanic acid (a fungal-derived inhibitor of β-lactamase enzymes) with β-lactam antibiotics. In response to the misuse of β-lactam antibiotics over an extended period, many bacterial strains have evolved to produce β-lactamase enzymes which cleave the β-lactam ring structure of these antibiotics, rendering them ineffective. Clavulanic acid is a weak β-lactam with negligible intrinsic antimicrobial activity on its own despite sharing a similar β-lactam ring with other β-lactam antibiotics. The similarity in chemical structure allows the molecule to bind β-lactamase irreversibly and act as an inhibitor of the enzyme. The antibiotic chemotherapy named Augmentin® (a combination of amoxicillin and potassium clavulanate) is formulated to take advantage of these synergistic combinational effects and has effectively repurposed β-lactam antibiotics for use against β-lactam resistant bacteria. The combination of β-lactam antibiotics such as amoxicillin with β-lactam inhibitors such as clavulanic acid is substantially more effective against *M. tuberculosis* than amoxicillin alone. Furthermore, clavulanic acid in combination with ampicillin, cephalothin, cephaloridine, or cefamandole is proven to act synergistically (reduced MIC and minimal bactericidal concentration) against β-lactamase-producing *S. aureus* and *Enterobacteriaceae*. There is enough evidence to suggest that the β-lactamase inhibitor may bind irreversibly, contributing to the overall efficacy of the antibiotic component of the combination [41].

Class II Syncretic Adjuvants

These adjuvants are increase antibiotic efficacy by altering host biology. The host defence systems within an infected organism offer valuable opportunities that can be exploited to improve the efficacy of antibiotics. For example, immunomodulatory antimicrobial peptides that function in a varied fashion in host immune systems have been proved to be synergistic with antibiotics and can even enhance antibiotic activity.

Syncretic class Ia antibiotic adjuvants		
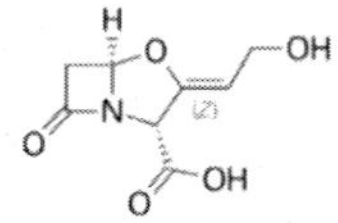	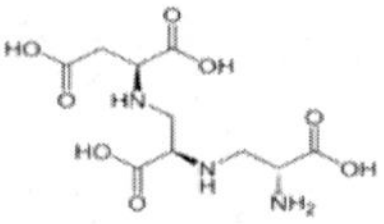	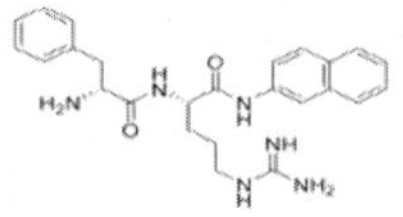
Clavulanic acid (FDA-approved) Serine- β-lactamase inhibitor [28]	Aspergillomarasmine A asmine A Metallo- β-lactamase inhibitor [29]	PaβN Efflux inhibitor [30]
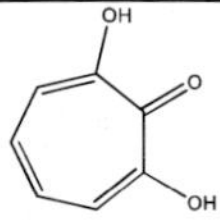	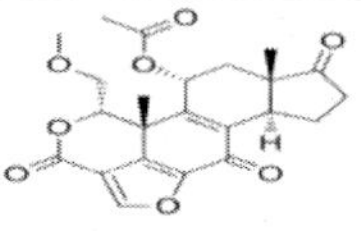	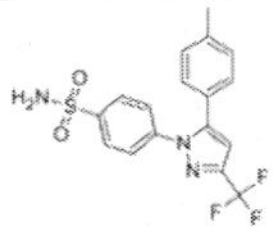
7-Hydroxytropolone Aminoglycoside adenyltransferase inhibitor [31]	Wortmannin Aminoglycoside kinase AP(2″) Inhibitor [32]	Celecoxib Efflux inhibitor [33]
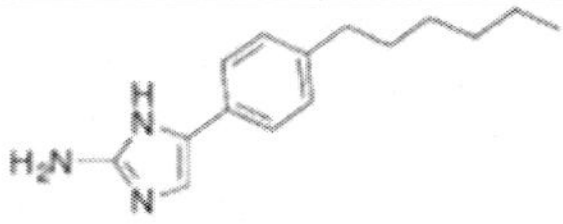	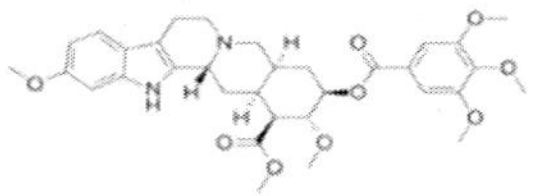	
2-aminoimidazole derivatives interferes with TCS signalling in A. baumannii to suppress colistin resistance [34]	Reserpine Efflux inhibitor [35]	
Syncretic class Ib antibiotic adjuvants		
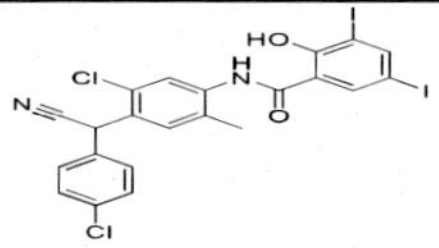	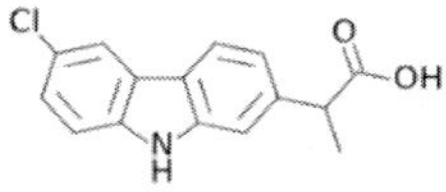	
Closantel Potentiates polymyxin B against Acinetobacter baumannii [35]	Carprofen Potentiates doxycycline against methicillinresistant Staphylococcus Pseudintermedius [36]	
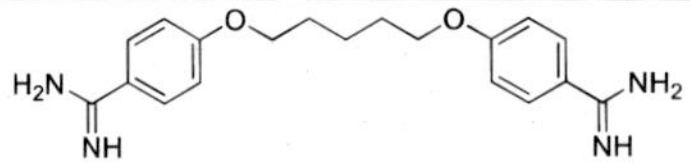	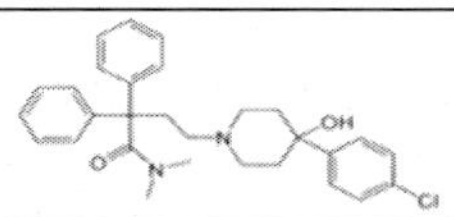	
Pentamidine Lipopolysaccharide disrupter that potentiates antibiotics against Gram- positive bacteria (for example, rifampin and novobiocin) in E. coli [37]	Loperamide Potentiates tetracycline antibiotics by perturbation of the cell membrane proton motive force against several Gram-negative bacteria [38]	
Hamamelitannin (quorum sensing inhibitor increases the sensitivity of S. aureus biofilms to several classes of 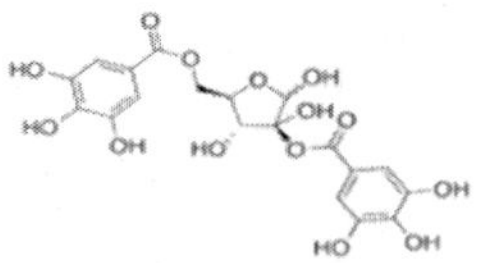 antibiotics)		

Figure 5. Some representative examples of syncretic antibiotic adjuvants.

The direct target of innate immunity with small molecule also demonstrated a significant advantage against resistance. For example, extracts of streptazolin, a natural product, activated the production of nuclear factor- κB over the phosphatidylinositide signalling pathway, in addition to the release of anti-infective cytokines, capable of stimulating macrophage killing of *S. pneumoniae.*

The therapeutic modulators of the host microenvironment such as alteration of the growth environment such as like nutrient availability for a pathogen radically change the sensitivity to antibiotics. Recent technology such as of CRISPR–Cas9 genetic screening technology has identified hundreds of host genetic loci that contribute to infection resistance, should enable the discovery of many Class II adjuvants that mimic the effect of host resistance determinants and serve as an untapped vista to explore new opportunities for future [42, 43].

Syncretic combinations of antibiotics with nonantibiotic adjuvants offer a very promising area for antibiotic discovery and development. In an era when new antibiotic innovation is at a nadir, reinvigorating our existing antibiotic drug classes provides an excellent opportunity to extend the life of well- researched and clinically validated drugs. The outstanding success of the Class Ia adjuvants that block serine- β-lactamase activity is evidence that this strategy is worthy of continued exploration. Other apparent targets for Class Ia adjuvants include metallo-β-lactamases, ribosome methyltransferases that confer near- pan resistance to aminoglycosides, large ribosome subunit methyltransferases such as Erm and Cfr that confer resistance to macrolide and oxazolidinone antibiotics, respectively, and broad-spectrum efflux inhibitors, in particular of the RND class that predominates in Gram-negative pathogens. Class Ib and Class II adjuvants are not yet in clinical development, but given the success of β- lactamase inhibitors and the growing clinical need, there is excellent opportunity to pursue these as well. In particular, Class Ib adjuvants could be targeted to antibiotics that in monotherapy have failed clinical trials owing to the emergence of resistance. Much is already known about such compounds, and these may offer highly suitable scaffolds for combination therapies. Although Class 1b adjuvants are unlikely to overcome serious deficiencies that have led to the triage of such antibiotic candidates, such as unmanageable toxicity and chemical or metabolic instability, clever deployment of Class II adjuvants may enable resurrection of abandoned antibiotics.

Coalism Combination

The term coalistic pairs used for combinations of non-antibiotic inhibitors but synthetically lethal gene functions. In Figure 3, X and Y are compounds without antibiotic activity served as coalistic pairs. Recent evidence suggested that the higher-order combinations of three or more compounds in principle can mimic more complex genetic interactions. Such combinations of compounds may lessen the frequency of resistance because inhibitors that target non- essential gene lack intrinsic antibiotic activity as single agents and thus afford less opportunity for maintenance in the absence of selection for resistance in microbial populations. These strategies offer new routes to narrow-spectrum antibiotics. Such therapies are increasingly seen as advantageous over broadspectrum drugs that select for resistance in multiple genera and indiscriminately damage the microbiome [42, 43].

The development of coalistic combinations represents an unexplored frontier that is now in principle accessible through systems biology and computational approaches. We believe that higher-order combinations of three or more compounds will be needed. Although much exploratory research and preclinical development are needed, the vast landscape of genetic interactions may well be exploited in a narrow- spectrum species-specific fashion. The myriad examples in nature of combinational strategies to combat pathogens should inspire a diversity of empirical and computational approaches. In particular, the plethora of ternary genetic interactions recently described in yeast suggests that myriad higher-order combinations of compounds that mimic these interactions await discovery. These arguments resonate with the observation that effective antibiotics in nature have evolved to be impervious to resistance by activity against multiple targets.

A COMBINATION APPROACH IN NATURE

In wild-type, bacterial strain phenomenon of antibiotic resistance is very common. It is reported in the literature that some soil-derived actinomycetes, found to resistant against fifteen different antibiotic classes. Such urbane resistance raises a critical question: Why do bacteria still effective and produces antibiotics? One possible reason is that bacteria usually adopted a combination approach in the natural environment. A quite instructive example of a natural combination is S.clavuligerus which produces a combination of the beta-lactam antibiotic cephamycin and the beta-lactamase inhibitor clavulanate, which are encoded by adjacent biosynthetic gene clusters, probably because they are transmitted horizontally together as part of a large mobile element.

The other example of natural combination therapy is *M. carbonacea*, (a soil actinomycete), which produces three natural products that are completely unrelated in chemical structure, each of which binds the ribosome at a distinct site. Selective pressure has presumably led to the maintenance of all three gene clusters in *M. carbonacea*, resulted in each of these are required for this strain to compete effectively with other species [44].

CHALLENGES FOR COMBINATION THERAPY

From the drug discovery point of view, this combined drug therapy has the advantage that it is not necessary to expend effort in the challenging and expensive identification of new targets that are essential for bacterial survival. The Adjuvant-antibiotic combination approach offers a more attractive option for the treatment of drug-resistant bacterial infections than the use of multiple antibiotics.

The principal challenge to the successful deployment of combination strategies as new medicines lies in the complex pharmacology of antibiotic action. Achieving the correct therapeutic levels and duration for a single antibiotic agent is already exceedingly difficult. Reaching these goals for two compounds that must be matched in terms of their pharmacokinetics and dynamics to maintain synergy considerably increases the complexity of drug development. For congruous pairs, if no historical data are available, clinical trials may need to include single agents in distinct arms of the trial. These concerns may not apply in the case of syncretic combinations, but formulation and administration may be complicated. Of course, before clinical trials, toxicology of each component and the combination must also be thoroughly investigated in case there are unexpected drug-drug interactions. The complexity rises for higher-order combinations. One solution is the synthesis of single-agent hybrids that combine, in one molecule, the bioactive domains of each component. Such hybrids can suppress resistance and even gain new modes of action.

The future of combination therapies can be summarised by discussing some key questions that reasonably anticipated its influence on the drug discovery pathway. Firstly, can antibiotics that have been freeze for having a high intrinsic resistance rate be rejuvenated as components of new combination therapies? Several interesting antibiotic candidates have been deferred due to a high intrinsic resistance rate. By rekindle these molecules as cocktail therapy, a cluster of new antibiotic scaffolds with improved resistance profiles could be introduced into clinical use [45]. What strategy used by microorganisms to design their combination therapies should be properly addressed? The ability to naturally produced antibiotics to potently inhibit a target; could provide us slew lessons about the design principles for antibiotics. It is worth mentioning that microbes have found a means for their antibiotics unaffected from widespread resistance under the same evolutionary pressures. we should pay keen attention to the lessons given by microbes for the design strategy of combination therapies.

CONCLUSION

The use of antimicrobials in clinical practice is a recent development in history compared to the emergence of bacterial organisms on our planet.

Therefore, the development of antibiotic resistance should be viewed as a "normal" adaptive response and a clear manifestation of Darwinian's principles of evolution. Arguably, the implementation of antimicrobial therapy in clinical practice has been one of the most successful advances of modern medicine, paving the way for complex and highly sophisticated medical interventions that have allowed to significantly prolong the living span of the population around the globe. To survive, bacteria, in a process likely pressed by the increased use of antimicrobials in clinical practice, have developed complex and creative strategies to circumvent the antibiotic attack. Antibiotic resistance has rapidly evolved in the last few decades to become now one of the greatest public health threats of the 21st century. Indeed, infections that are untreatable due to multidrug resistance of the infected organism have become more common in clinical settings. This dire scenario has been worsened by a shortage of research and development on antibiotics. The "golden" pipeline of antibiotic discovery (the 1960s and 70s) rapidly dried up as the identification of new compounds became more challenging. Big pharma concentrated their efforts on other more profitable and rewarding areas, leaving the wave of resistance to grow unabated. If we are to tackle this problem, efforts on research and development need to be heavily increased and supported. A complete understanding of the mechanisms by which bacteria become resistant to antibiotics is of paramount importance to design novel strategies to counter the resistance threat. We require developing antibiotics with the understanding that the microorganism will respond to them and resistance will develop (an evolutionary fact). Therefore, efforts to develop antibiotics and study mechanisms of resistance should be continuous, resilient and steady. This is likely to be a long haul "war" against living entities with a major ability to adapt and survive. In conclusion, every potential avenue must be utilized if we are to overcome this crisis. Although the development of new antibiotics remains crucial, and improvements in antibiotic stewardship are essential, the adoption of complementary approaches such as the development of adjuvants that combat antibiotic resistance represents a powerful and underexploited weapon.

ACKNOWLEDGMENT

Financial assistance in the form of UGC start-UP (F.30-461/2019 (BSR) to Anuradha Singh by the UGC, New Delhi is sincerely acknowledged.

REFERENCES

[1] Bertheim, A. (2015). Pioneers in Antimicrobial Chemotherapy. *J Assoc Physicians India*, *63*, 90.

[2] Davies, J. E., & Behroozian, S. (2020). An ancient solution to a modern problem. *Molecular Microbiology*, *113*(3), 546-549.

[3] Brown, E. D., & Wright, G. D. (2016). Antibacterial drug discovery in the resistance era. *Nature*, *529*(7586), 336-343.

[4] Singh, S. B., Young, K., & Silver, L. L. (2017). What is an "ideal" antibiotic? Discovery challenges and path forward. *Biochemical pharmacology*, *133*, 63-73.

[5] https://www.who.int/news-room/fact-sheets/detail/antimicrobial-resistance.

[6] Chokshi, A., Sifri, Z., Cennimo, D., & Horng, H. (2019). Global contributors to antibiotic resistance. *Journal of global infectious diseases*, *11*(1), 36.

[7] Santajit, S., & Indrawattana, N. (2016). Mechanisms of antimicrobial resistance in ESKAPE pathogens. *BioMed research international*, *2016*.

[8] Carlet, J., Jarlier, V., Harbarth, S., Voss, A., Goossens, H., & Pittet, D. (2012). *Ready for a world without antibiotics? The pensières antibiotic resistance call to action*.

[9] https://www.who.int/news-room/detail/29-04-2019-new-report-calls-for-urgent-action-to-avert-antimicrobial-resistance-crisis.

[10] Holmes, A. H., Moore, L. S., Sundsfjord, A., Steinbakk, M., Regmi, S., Karkey, A., ... & Piddock, L. J. (2016). Understanding the

mechanisms and drivers of antimicrobial resistance. *The Lancet*, *387*(10014), 176-187.

[11] Aminov, R. I. (2009). The role of antibiotics and antibiotic resistance in nature. *Environmental Microbiology*, *11*(12), 2970-2988.

[12] Masterton, R. G., Bassetti, M., Chastre, J., MacDonald, A. G., Rello, J., Seaton, R. A., ... & West, P. (2019). *Valuing antibiotics: The role of the hospital clinician.*

[13] Lewis, K. (2020). The Science of Antibiotic Discovery. *Cell.*

[14] Aslam, B., Wang, W., Arshad, M. I., Khurshid, M., Muzammil, S., Rasool, M. H., ... & Salamat, M. K. F. (2018). Antibiotic resistance: a rundown of a global crisis. *Infection and drug resistance*, 11, 1645.

[15] Ouellette, M., & Bhattacharya, A. (2020). Exploiting antimicrobial resistance. *EMBO reports.*

[16] Douafer, H., Andrieu, V., Phanstiel IV, O., & Brunel, J. M. (2019). Antibiotic adjuvants: make antibiotics great again!. *Journal of medicinal chemistry*, *62*(19), 8665-8681.

[17] Sciarretta, K., Røttingen, J. A., Opalska, A., Van Hengel, A. J., & Larsen, J. (2016). Economic incentives for antibacterial drug development: literature review and considerations from the Transatlantic Task Force on Antimicrobial Resistance. *Clinical Infectious Diseases*, *63*(11), 1470-1474.

[18] Wright, G. D. (2016). Antibiotic adjuvants: rescuing antibiotics from resistance. *Trends in microbiology*, *24*(11), 862-871.

[19] Munita, J. M., & Arias, C. A. (2016). Mechanisms of antibiotic resistance. *Virulence mechanisms of bacterial pathogens*, 481-511.

[20] Singh, A., & Singh, R. K. (2018). Human Immunodeficiency Virus Reverse Transcriptase (HIV-RT): Structural Implications for Drug Development. In Bharati, S. L., & Chaurasia, P. K. (Ed.), *Research Advancements in Pharmaceutical, Nutritional, and Industrial Enzymology* (pp. 100-127). IGI Global. http://doi:10.4018/978-1-5225-5237-6.ch005.

[21] Worthington, R. J., & Melander, C. (2013). Combination approaches to combat multidrug-resistant bacteria. *Trends in Biotechnology*, *31*(3), 177-184.

[22] Coates, A. R., Hu, Y., Holt, J., & Yey, P. (2020). Antibiotic combination therapy against resistant bacterial infections: synergy, rejuvenation and resistance reduction. *Expert Review of Anti-infective Therapy.*

[23] Brown, D. (2015). Antibiotic resistance breakers: can repurposed drugs fill the antibiotic discovery void?. *Nature reviews Drug discovery*, *14*(12), 821-832.

[24] Hand, D. J. (2000). Synergy in drug combinations in *Data Analysis* (pp. 471-475). Springer, Berlin, Heidelberg.

[25] Zhou, A., Kang, T. M., Yuan, J., Beppler, C., Nguyen, C., Mao, Z., ... & Miller, J. H. (2015). Synergistic interactions of vancomycin with different antibiotics against Escherichia coli: trimethoprim and nitrofurantoin display strong synergies with vancomycin against wild-type E. coli. *Antimicrobial agents and chemotherapy*, *59*(1), 276-281.

[26] Cheesman, M. J., Ilanko, A., Blonk, B., & Cock, I. E. (2017). Developing new antimicrobial therapies: are synergistic combinations of plant extracts/compounds with conventional antibiotics the solution?. *Pharmacognosy reviews*, *11*(22), 57.

[27] Tyers, M., & Wright, G. D. (2019). Drug combinations: a strategy to extend the life of antibiotics in the 21st century. *Nature Reviews Microbiology*, *17*(3), 141-155.

[28] Brown, A. G., Butterworth, D., Cole, M., Hanscomb, G., Hood, J. D., Reading, C., & Rolinson, G. N. (1976). Naturally-occurring β-lactamase inhibitors with antibacterial activity. *The Journal of antibiotics*, *29*(6), 668-669.

[29] King, A. M., Reid-Yu, S. A., Wang, W., King, D. T., De Pascale, G., Strynadka, N. C., ... & Wright, G. D. (2014). Aspergillomarasmine A overcomes metallo-β-lactamase antibiotic resistance. *Nature*, 510(7506), 503-506.

[30] Lomovskaya, O., Warren, M. S., Lee, A., Galazzo, J., Fronko, R., Lee, M. A. Y., ... & Leger, R. (2001). Identification and characterization of inhibitors of multidrug resistance efflux pumps in *Pseudomonas aeruginosa*: novel agents for combination therapy. *Antimicrobial agents and chemotherapy*, *45*(1), 105-116.

[31] Allen, N. E., Alborn, W. E., Hobbs, J. N., & Kirst, H. A. (1982). 7-Hydroxytropolone: an inhibitor of aminoglycoside-2"-O-adenylyltransferase. *Antimicrobial Agents and Chemotherapy*, *22*(5), 824-831.

[32] Boehr, D. D., Lane, W. S., & Wright, G. D. (2001). Active site labelling of the gentamicin resistance enzyme AAC (6′)-APH (2 ″) by the lipid kinase inhibitor wortmannin. *Chemistry & Biology*, *8*(8), 791-800.

[33] Kalle, A. M., & Rizvi, A. (2011). Inhibition of bacterial multidrug resistance by celecoxib, a cyclooxygenase-2 inhibitor. *Antimicrobial agents and chemotherapy*, *55*(1), 439-442.

[34] Harris, T. L., Worthington, R. J., Hittle, L. E., Zurawski, D. V., Ernst, R. K., & Melander, C. (2014). Small molecule downregulation of PmrAB reverses lipid A modification and breaks colistin resistance. *ACS chemical biology*, *9*(1), 122-127.

[35] Stermitz, F. R., Lorenz, P., Tawara, J. N., Zenewicz, L. A., & Lewis, K. (2000). Synergy in a medicinal plant: antimicrobial action of berberine potentiated by 5′-methoxyhydnocarpin, a multidrug pump inhibitor. *Proceedings of the National Academy of Sciences*, *97*(4), 1433-1437.

[36] Tran, T. B., Cheah, S. E., Heidi, H. Y., Bergen, P. J., Nation, R. L., Creek, D. J., ... & Velkov, T. (2016). Anthelmintic closantel enhances bacterial killing of polymyxin B against multidrug-resistant Acinetobacter baumannii. *The Journal of antibiotics*, *69*(6), 415-421.

[37] Brochmann, R. P., Helmfrid, A., Jana, B., Magnowska, Z., & Guardabassi, L. (2016). Antimicrobial synergy between carprofen and doxycycline against methicillin-resistant Staphylococcus pseudintermedius ST71. *BMC veterinary research*, *12*(1), 126.

[38] Stokes, J. M., MacNair, C. R., Ilyas, B., French, S., Côté, J. P., Bouwman, C., ... & Brown, E. D. (2017). Pentamidine sensitizes Gram-negative pathogens to antibiotics and overcomes acquired colistin resistance. *Nature microbiology*, *2*(5), 1-8.

[39] Ejim, L., Farha, M. A., Falconer, S. B., Wildenhain, J., Coombes, B. K., Tyers, M., ... & Wright, G. D. (2011). Combinations of antibiotics

and nonantibiotic drugs enhance antimicrobial efficacy. *Nature chemical biology*, *7*(6), 348.

[40] Brackman, G., Breyne, K., De Rycke, R., Vermote, A., Van Nieuwerburgh, F., Meyer, E., ... & Coenye, T. (2016). The quorum sensing inhibitor hamamelitannin increases antibiotic susceptibility of Staphylococcus aureus biofilms by affecting peptidoglycan biosynthesis and eDNA release. *Scientific reports*, *6*, 20321.

[41] Neu HC, Fu KP. Clavulanic acid, a novel inhibitor of β-lactamases. *Antimicrob Agents Chemother*. 1978;14:650–5.

[42] Melander, R. J., & Melander, C. (2017). The challenge of overcoming antibiotic resistance: an adjuvant approach?. *ACS infectious diseases*, *3*(8), 559-563.

[43] Brown, D. (2015). Antibiotic resistance breakers: can repurposed drugs fill the antibiotic discovery void?. *Nature reviews Drug discovery*, *14*(12), 821-832.

[44] Fischbach, M. A. (2011). Combination therapies for combating antimicrobial resistance. *Current opinion in microbiology*, 14(5), 519-523.

[45] Coates, A. (2019). The future of antibiotics lies in combination treatments. *Future Drug Discovery*, 1(1), FDD5.

BIOGRAPHICAL SKETCH

Anuradha Singh

Affiliation: University of Allahabad

Education: M Sc, PhD

Business Address: Department of Chemistry
Sadanlal Sanwaldas Khanna Girls' Degree College
(A Constituent College of the University of Allahabad)
Prayagraj – 211003, INDIA
Mob. 8400055477

Email: anuradha_au@rediffmail.com

Research and Professional Experience:

Dr Anuradha Singh (Assistant professor in Department of Chemistry, S S Khanna Girls' Degree College, University of Allahabad) recently completed a research project entitled "Design and Development of bi-functional Reverse Transcriptase Inhibitors of HIV Replication and Their Enhanced Cellular Uptake Using Myristoylated Derivatives" funded by the Department of Science and Technology (DST), New Delhi as Principal Investigator. Presently she is working as Principal Investigator in a start-up project funded by University Grants Commission, New Delhi. Her current research interest is focused around developing multifunctional anti-viral and antimicrobial agent. She is well versed in synthetic chemistry, computer-aided drug design and other techniques used in the field of medicinal chemistry. Dr Singh has published more than 20 research papers/book chapters, in addition to several articles, in journals/books of national and international repute.

Professional Appointments: Assistant Professor (2018)

Honours:
Swarna Jayanti Puraskar, 2014, 84th Annual session of NASI, Jodhpur
HIV symposium fellowship, 2012, Chennai
Late M.H. Mahalaha memorial Gold medal for best presentation at the seminar, 2005, RDVV, Jabalpur

Publications from the Last 3 Years:

Research articles:

1) M. Yadav, R. Srivastava, F. Naaz, Anuradha Singh, R. Verma and R. K. Singh (2019). *In silico* studies on new oxathiadizoles as potential anti-HIV agents, *Gene Reports*, 14, 87-93.

2) Ritika Srivastava, Sunil K. Gupta, Farha Naaz, Anuradha Singh, Vishal Kumar Singh, Rajesh Verma, Nidhi Singh and Ramendra K. Singh (2018). Synthesis, antibacterial activity, synergistic effect, cytotoxicity docking and molecular dynamics of benzimidazole analogues, *Computational Biology and Chemistry*, 76, 1-16.
3) Farha Naaz, Ritika Srivastava, Anuradha Singh, Nidhi Singh, Rajesh Verma, Vishal Kumar Singh and Ramendra K. Singh (2018). Molecular modelling, synthesis, antibacterial and cytotoxicity evaluation of sulfonamide derivatives of benzimidazole, indazole, benzothiazole and thiazole, *Bioorganic and Medicinal Chemistry*, 26, 3414-3428.
4) Anuradha Singh, V. K. Singh, R. Verma and Ramendra K. Singh (2018) *In silico* studies on *N*-(pyridine-*2-yl*) thiobenzamides as NNRTIs against wild and mutant HIV-1 Strains. *Philippine Journal of Science*, 147 (1), 37-46.
5) Anuradha Singh, Ritika Srivastava, and Ramendra K. Singh (2017). Design, synthesis, and antibacterial activities of novel heterocyclic arylsulphonamide derivatives, *Interdisciplinary Sciences Computational Life Sciences*, doi 10.1007/s12539-016-0207-2.
6) M. Yadav, Anuradha Singh and Ramendra K. Singh (2017) Docking studies on novel bisphenylbenzimidazoles (BPBIs) as non-nucleoside inhibitors of HIV-1 reverse transcriptase. *Indian Journal of Chemistry (B)*, 56B, 714-723.
7) Nidhi Singh, Ritika Srivastava, Anuradha Singh and Ramendra K. Singh (2016). Synthesis & Photophysical studies on naphthalimide derived fluorophores as markers in drug delivery, *Journal of Fluorescence*, 26, 1431- 1438.
8) Anuradha Singh, Madhu Yadav, Ritika Srivastava, Nidhi Singh, Rajinder Kaur, Satish K. Gupta and Ramendra K. Singh (2016). Design and anti-HIV activity of arylsulphonamides as non-nucleoside reverse transcriptase inhibitors, *Medicinal Chemistry Research*, 25, 2842 - 2859.

Chapters in books

1) Anuradha Singh (2020). Pharmacological Properties of Curcumin: Solid Gold or Just Pyrite?. In *Advanced Pharmacological Uses of Medicinal Plants and Natural Products* (pp. 235-248). IGI Global. doi: 10.4018/978-1-7998-2094-9.ch012.
2) Anuradha Singh (2019). Molecular docking: A computational tool for structure based drug discovery. In: *Advances in Chemical and Applied Sciences* (pp. 170-181). First print publications. ISBN: 978-93-88018-17-3.
3) Anuradha Singh, and Ramendra K. Singh (2018). Human Immunodeficiency Virus Reverse Transcriptase (HIV-RT): Structural Implications for Drug Development. In *Research Advancements in Pharmaceutical, Nutritional, and Industrial Enzymology* (pp. 100-127). IGI Global. doi: 10.4018/978-1-5225-5237-6.ch005.

INDEX

D

E

F

G

H

I

K

L

M

N

O

P

R

S